MICRO-ART

ART IMAGES IN A HIDDEN WORLD

text and photographs by Lewis R. Wolberg M.D.

WITH A PREFACE BY BRIAN O'DOHERTY

HARRY N. ABRAMS, INC., PUBLISHERS • NEW YORK

MICRO-ART

ART IMAGES IN A HIDDEN WORLD

Book design by Robin Fox

Standard Book Number: 8109–0302–4

Library of Congress Catalogue Card Number: 78–119623

Printed and bound in Japan

CONTENTS

63. Airborne seed
300x

64. Fungus: *Saccharomyces*
3,000x Polarized light

65. Cucumber wilt
500x

66, 67. Green algae: *Laminaria* blade with sporangia
375x Dark field

68. Desmids
750x

69. Virus
160,000x Taken with an electron microscope

70. White oak *(Quercus alba)*
300x Polarized light

71. White oak *(Quercus alba)*
300x

72. Diatom: *Stauroneis* (bacteria in background)
1,690x

73. Leaf skeleton
225x Dark field

74. Fungi: *Ascomycetes*
700x
Fruit cells *(microphaera ascocarps)*

75. Eastern hemlock *(Tsuga canadensis)*
230x

76, 77. Pine stem
320x

78. Dicotyledonous plant: *Tilia* stem
180x Dark field

79, 80. Beech *(Fagus grandifolia)*
490x

81. Sponge: *Sycon ciliatum*
1,000x Polarized light

82, 83. Bamboo *(Bambusa)*
600x Polarized light

83. Bamboo *(Bambusa)*
600x

84. Green algae: *Fucus,* male conceptacle
700x

85. Liverwort
3,000x Polarized light

86. *Volvox*
525x

87. Diatoms
750x

88. Potato scab
750x Polarized light

89. Fungus: potato scab
190x Polarized light

90, 91. Pine *(Pinus banksiona)*
1200x

92. Green radish root
300x Polarized light

93. Moss: *Mnium*
300x
Female sexual organs *(archegonia)*

94. Leaf stalk of garden radish
500x

95. Leaf stalk of garden radish
565x Dark field

96. Green algae: *Vaucheria*
200x

ANIMAL 99

---

106. Esophagus
250x

107. Protozoa: *Tetrahymena pyriformis*
750x

108. Cerebellum
790x

109. Umbilical cord
200x

110, 111. Angora rabbit hair
1,100x Polarized light

112. *Hydra* nematocysts
1,500x

113. Protozoan: *Radiolarian*
2,300x

114. Mouse embryo
300x

115. Cow hair
600x Polarized light

116. Protozoa: *Radiolaria* amidst bacteria
1,100x

117. Protozoa: *Paramecia*
1,400x Polarized light

right: 118. External ear
300x

left: 118. Protozoan: fossil *Radiolarian*
800x

119. Protozoan: *Foraminiferidan*
1,900x
Polarized light

120, 121. Pituitary
1,150x

122. Protozoan: *Radiolarian*
750x

123. Spermatic cord
235x

124. Protozoa: *Opercularia* colony
580x

125. Mouth parts of male mosquito
175x

125, 126. Aphid
100x

127. Butterfly wing
1,050x Polarized light

128. Butterfly wing
140x Dark field

129. Butterfly wing
1,260x

130. Amphibian ovary: *taricha*
180x

131. Rabbit esophagus
900x

132, 133. Epididymis
460x

134. Lip
1,200x

135. Protozoan: *Stentor*
1,875x

136. Bladder
280x

137. Developing foot of embryo
250x

138. Protozoan: *Radiolarian*
2,000x

139. Shad ovary
400x

right: 140. Flatworm: planaria
580x

left: 140. Flatworm: planaria
230x

141. Frog testes
600x

142, 143. Bat hair
460x Polarized light

144. Fish scale: ctenoid
200x

145. Internal ear of guinea pig: organ of Corti
250x

146. Sciatic nerve
250x

147. Vagina
900x

148. Ureter
800x

149, 150. Radule in digestive system of snail
570x

151. Fish scale: placoid
225x

152. Pigeon esophagus
800x

153. Chinese liver fluke *(Clonorchis sinensis)*
190x

154, 155. Sea anemone: *Metridium*
460x

156. Hydra
210x

157. Pylorus of stomach
300x

158. Trout stomach
700x

159. Protozoan: *Foraminiferidan*
1,875x

160. Protozoan: *Foraminiferidan*
1,900x

161. Roundworm: *Ascaris*
600x
Cellular cleavage showing mitosis

162, 163. Urethra
750x

163. Protozoan: *Amoeba proteus*
790x

164. Developing tooth
260x

165. Monkey brain
175x

166. Sponge spicules
300x Polarized light

167, 168. Turtle tongue
880x

MINERAL 171

176. **Nickel sulfate**
300x

177. **Tremolite**
225x

178. **Staurolite**
300x

179. **Andalusite**
240x **Polarized light**

180. **Opal**
300x **Polarized light**

181. **Study of graphite**
1,300x **Interference contrast**
Courtesy of Edith M. Raviola, General Electric Co., Research and Development Center, Schenectady, N.Y.

182. **Silicon**
500x **Polarized light**

183. **Silicon wafer**
400x

184, 185. **Cubic crystalline structure of etched tungsten**
20,600x **Taken with an electron microscope**
Courtesy of Clark D. Smith, Space Division, The Boeing Co., Seattle, Wash.

186. **Cholesterol ester**
200x **Polarized light**

187. **Amino acid: glycine**
740x **Polarized light**

188. **Detail of a surface coating on titanium metal**
4,500x **Taken with an electron microscope**
Courtesy of Clark D. Smith, Space Division, The Boeing Co., Seattle, Wash.

189. **Sugar**
200x **Polarized light**

190. **Cholesterol ester**
500x **Polarized light**

191. **Cholesterol ester**
375x **Polarized light**

192, 193. **Maraging steel**
9,000x
Courtesy of Frank D. Walsh, Manufacturing Research and Development, The Boeing Co., Seattle, Wash

Magnesium alloy
Courtesy of Frank D. Walsh, Manufacturing Research and Development, The Boeing Co., Seattle, Wash.

194. **Nickel-aluminum bronze, transverse cracks during tensile test** **32,000x**
Courtesy of W. Gifkins, British Welding Research Association, Cambridge, England.

195. **Amino acid: *DL* norleucine**
200x **Polarized light**

196. **Cast steel (pearlite)**
100,000x **Taken with an electron microscope**
Courtesy of Ulla Fuchs and Ursula L. Schaaber, Institut für Harterei-Technik, Bremen, Germany.

197. **Field ion micrograph of a nickel molybdenum alloy to study composition at an atomic level**
3,000,000x
Courtesy of R. W. Newman, B. G. LeFevre, J. J. Hren, University of Florida, Gainesville, Fla.

198. **Dayton iron meteorite**
475x **Polarized light with interference plate**
Courtesy of Harvey Yakowitz, National Bureau of Standards, Wash. D.C., and J. I. Goldstein, Goddard Space Flight Center Greenbelt, Md.

199. **Amino acid: *DL* leucine**
800x **Polarized light**

200. **High carbon steel (martensite plate)**
25,000x **Taken with an electron microscope**
Courtesy of Arlan D. Benscoter, Bethlehem Steel Corp., Bethlehem, Pa.

201. **Copper-cadmium-zinc alloy**
16,000x
Courtesy of Karin Baer, Max Planck-Institut für Metallforschung, Stuttgart, Germany.

202. **Epsom salts**
700x **Polarized light**

203. **Resorcinol doped with tartaric acid: variations in alignment of crystals**
60x **Polarized light**
Courtesy of Andrew S. Holik, General Electric Co., Research and Development Center, Schenectady, N.Y.

204. **Vesuvianite**
300x

205, 206. **Amino acid: *DL* isoleucine**
490x **Polarized light**

207, 208. **Chromium-nickel subjected to sulfidation**
14,900x
Courtesy of Robert R. Russell, Research and Development Center, General Electric Co., Schenectady, N.Y.

209. **Rochelle salts (potassium sodium tartrate)**
200x **Polarized light**

210. **Noibium (columbium) subjected to vacuum degassing at 2,300°C under strong pressure**
8,800x **Taken with an electron microscope**
Courtesy of Robert W. Meyerhoff, Union Carbide Corp., Indianapolis, Ind.

211. **Nepheline**
180x **Polarized light**

212, 213. **Clont (1-hydroxy ethyl-2-methyl-5-nitroimidazol)**
575x **Polarized light**
Courtesy of Carl Zeiss, Inc., New York.

214. **Crystal growth in the oxidation of lead**
42,600x **Taken with an electron microscope**
Courtesy of Edward F. Koch, General Electric Co., Schenectady, N.Y.

top: **215. Copper-zinc alloy**
500x
Courtesy of Buehler Ltd., Evanston, Ill.

bottom: **215. Molybdenum-titanium alloy**
250x
Courtesy of Buehler Ltd., Evanston, Ill.

216. Graphite subjected to tension to induce a fatigue fracture
175,000x Taken with an electron microscope
Courtesy of Floyd B. Brate, Jr., General Electric Co., Aircraft Engine Group, Evandale, Ohio.

217. Titanium sheet subjected to nitrogen at a high temperature
600x Polarized light with interference plate
Courtesy of Harvey Yakowitz, National Bureau of Standards, Wash. D.C.

218. Dendroidal carbide extracted from an intergranular fracture of stainless steel
125,000x Taken with an electron microscope
Courtesy of Visvaldis Biss, Climax Molybdenum Company of Michigan, Ann Arbor, Mich.

219. Prehnite
500x Polarized light

220. Pyrite, magnolite, hemolite
150x

221. Pyrite
210x

222, 223. Microstructure of cast 250 grade maraging steel
30,000x
Courtesy of Dennis Ŝ. Crowther, Rudy B. Spanholtz, and Fred Winchell, Martin Marietta Corp., Orlando, Fla.

224. Silicon wafer
225x

225. Rochelle salts (potassium sodium tartrate)
190x Polarized light

226. Amino acid: *DL* norleucine
240x Polarized light

227. Potassium ferricyanide
1,200x Polarized light

228. Etched surface of a fracture in magnesia-alumina specimen
50,000x Taken with an electron microscope
Courtesy of Clark D. Smith, Space Division, The Boeing Co., Seattle, Wash.

229. Iron-nickel alloy
760x Polarized light, stain etching
Courtesy of Arlan D. Benscoter, Bethlehem Steel Corp., Bethlehem, Pa.

230. Uranium-iron; dendrite along a central axis
12,400x Taken with an electron microscope
Courtesy of I. D. Miller, General Electric Co., Nuclear Materials and Propulsion Operation, Cincinnati, Ohio.

231. Orthoclase
750x Polarized light

left: **232. Titanium alloy**
Co., Seattle, Wash.
375x
Courtesy of Frank D. Walsh, Manufacturing Research and Development, The Boeing Co., Seattle, Wash.

right: **232. Titanium alloy**
250x
Courtesy of Frank D. Walsh, Manufacturing Research and Development, The Boeing Co., Seattle, Wash.

left: **233. Titanium base metal**
435x
Courtesy of Frank D. Walsh, Manufacturing Research and Development, The Boeing Co., Seattle, Wash.

right: **233. Titanium alloy**
500x
Courtesy of Frank D. Walsh, Manufacturing Research and Development, The Boeing Co., Seattle, Wash.

234. Hypo (sodium thiosulfate)
700x Polarized light

235. Titanium brazed with copper foil
1,900x
Courtesy of Frank D. Walsh, Manufacturing Research and Development, The Boeing Co., Seattle, Wash.

236. Amino acid: *DL* tyrosine
300x Polarized light

237. Amino acid: glycine
200x

238,239. Gel-type plastic material: liquid crystal (fluor chemical) 3,600x Polarized light
Courtesy of Carl Zeiss, Inc., New York.

240. Cholesterol ester
500x Polarized light

241. Rochelle salts (potassium sodium tartrate)
200x Polarized light

242. Hypo (sodium thiosulfate)
700x

243. Potassium ferricyanide
375x Polarized light

244. Nickel sulfate
200x Polarized light

245. Rusted surface of a broken tool steel part
125,000x Taken with an electron microscope
Courtesy of Clark D. Smith, Space Division, The Boeing Co., Seattle, Wash.

246, 247. Martensite (the hardest and most brittle condition of steel) in an iron nickel alloy
11,600x Polarized light, stain etching
Courtesy of Arlan D. Benscoter, Bethlehem Steel Corp., Bethlehem, Pa.

248. Albite
700x Polarized light

249. Potassium chromium sulfate
230x Polarized light

PREFACE

"I dreamed that I floated at will in the great Ether," wrote Ralph Waldo Emerson, "and I saw this world floating not far off, but diminished to the size of an apple. Then an angel took it in his hand and brought it to me and said 'This must thou eat.' And I ate the world."

There are so many implications here that one can point to only a few most relevant to the theme of the miniature. Through reduction, the intractable resistances of objects are made pliable to thought. The result is a conceptual Utopia where thought and thing are mobilized in one easy and angelic swiftness. Thought suffers no resistance to its flow from the substance of its images, which are no longer detained by real weights and gravities.

The release from gravity (floating in the great Ether) is a similar Utopian idea. The body is ditched, but the senses are not. The eating of the apple—while obviously referring to the myth of knowledge—perhaps refers also to the discreet appetites provoked by smallness. The acute conceptual grasp of the tiny object has its visceral echo—as if *eating* it signified full possession, an abstract thought in an abstract stomach. This kind of psychological self-impregnation with the miniaturized object is a profound biological urge, as Baudelaire has pointed out with respect to toys.

The Ether itself signified a convergence of poetic and scientific ideas vital to the theme. The Ether now occupies a museum of obsolete scientific concepts—of substances that never existed but were made necessary by language and a mechanist idea of process. A history of such substances (including the Ether of communication, the phlogiston of combustion, the "dirt" of spontaneous generation) remains to be written. Not from the point of view of their obsolescence, but of their imaginative necessity. They represent the intrusion of Mind into inexplicable process. As delusions of Mind they recall the conjugations of "fancy" and idealist philosophy at the beginning of Romanticism.

With Romanticism the imaginative faculty becomes a lens through which changes in scale conjugate the idea of the self. Swift's reversals in size are not frightening at all; they are expository, not the locus of his own anxiety. Swift knew exactly what size *he* was, and his sentences move with aplomb across unperceived abysses that the Romantic imagination would inhabit. That imagination and Victorian science are the proper ancestors of this book. Ernst Haeckel's *Art Forms of Nature* (1899–1904) is a direct ancestor, "one of the first indications," writes Philip Ritterbush, "that the forms introduced to science by morphology and microscopy in the nineteenth century might have a broader esthetic appeal." Haeckel, who finally argued for souls in crystals, reminds us that a concept like the Ether—made up of science, imagination, spirit, and superstition—represents fairly accurately the context of feeling within which books like this one were received and to some extent still are. For Ether is a medium inaccessible to the senses. As the invisible part of the evidence, it was located between logic and blind faith. Through it physical effects arrived trailing the remnants of a supposed spiritual journey. Thus entangled with the invisible, material phenomena put a stress on the senses, invariably eliciting this metaphysical shudder—which in turn introduces us to the poetics of bourgeois wonder. This wonder is habitually inseparable from the microscopic. It is composed of a somewhat routine sense of miracle, informed by delectation and a quasi-religious sense of a higher order revealed in the microscopic. This optimism is the vernacular translation of Emerson's dream. At the same time, beneath its comfortable wonder, it calls up another, riskier history to which the microscope has also contributed—that of the body image.

II

"I am sensible of a certain doubleness," wrote Thoreau, "by which I stand as remote from myself as from another." This in its easy tone and clear reasonableness should not be assimilated into the literature of alienation. Consciousness becomes a mirror that turns the observing eye back on itself. There is a doubling of the senses, and this double eye is applied to the microscope, as elsewhere. To what is observed is added the process of observation.

This process inevitably involves those projections that are a form of self-identification. Sliding like some animalcule down the barrel of the microscope, the anthropomorphic image distributes itself across the hitherto invisible, "civilizing" the hidden landscape. Dimly it is understood that such observation of microworlds makes the body itself unfixed and labile. A change in scale indicates a change in value. Quantity becomes qualitative. The microscopic eye, focusing on the wafer of clarity between two blurred distances, returns a question on the creature who observes.

How pervasive the popular forms of that question are! Pascal's vision of a creature slid like a drawer into a cabinet of infinite scales, limited by his size and afraid to lose it, is translated into hosts of popular dreams—movies—where the Romantic and Gothick concerns of the nineteenth century are rehearsed. The horror movie has two main themes: organic mutation on the one hand, exaggeration in size on the other. The epiphanies provoked by hypertrophic monsters—animal or vegetable—are not so much awe at sheer size (a popular trace of the sublime), as at the human shrinkage it imposes. Terror of the monster is more truly terror at our insignificant personal size.

The classic of this anxiety, "The Incredible Shrinking Man" (1957), institutes with each change in size a change in values, creating new sets of monsters and problems until eventually an atom of Mind slips into infinitesimal anonymity. The microscope can be seen as an instrument of this anxiety. By diminishing the observer and querying his location in space, the self is lost in the general insurrection of objects that follows—making the body a monster to itself and introducing a new theme for literature.

The microscopic, like all extensions of the senses, initially confirms the senses but ends by questioning them. That doubled self-consciousness of which Thoreau spoke makes them artificial, that is, conditional. In the large theme of the body's displacement from the center of consciousness to a piece of fleshy luggage, the miniature and the microscopic have played distinct roles. (The miniature is that which can be held and touched and seen with the naked eye. The microscopic can be seen only with the aided eye.) The taste for the miniature—toys, dolls' houses, Japanese trees, Fabergé jewelry, and so on—is a form of Rococo reassurance, which is often tactile. The ecstasy over the tiny is partly due to the ease with which it can be destroyed. There is a destructive instinct held in abeyance in this connoisseurship, and miniaturization of this sort is the exercise of a form of power.

But the microscopic is deeply subversive because the eye is isolated, touch cannot follow (or can only be conceptualized and projected through the eye), and the body is locked outside. The possibility of scalelessness enters—the defining mark or index is gone. The observed terrain can shrink or expand until the microscopic and the macroscopic mimic each other endlessly. Moon photographs may represent a square foot or a hundred square miles. Information turns out to be endlessly equivocal. Each horizon of magnification seems to hold the promise of banishing some lurking insecurity but only rephrases and refines it. For each question solved another is raised. Each relay of enlargement, like Zeno's paradoxes of motion, ultimately encourages stasis.

This stasis may have something to do with the nature of the microscopic field: an isolated circle of light transfusing a depthless layer of material, usually immobile. This isolation, added to the clarity of the focused-on, is a further form of focus. Peripheral vision is cut out. The annular field is thus set at no particular distance, since there are no cues to distance. The field seems almost *within* the eye, a situation which, instead of encouraging intimacy, alienates the eye still further. The sharp circumference and clear focus might be expected to dramatize what is perceived, but though there is a quality of drama—isolation—the opposite is true. The equal clarity of the field tends to make it all-over and particulate. The circular edge is an area of low attention. The eye, apparently held rigid, retains an immense capacity to wander, though non-incident is as much focused as incident. Biology and pathology are modes of training the eye what to see, of identifying tiny gestalts that add up to a frame of reference. Sight, through the microscope, is literally invented—just as it is in other situations (for example, space flight) where localizing cues are absent. Returning to the everyday world, sight is eroded by suspicions that are hardly conscious.

The aesthetic of the microscopic vision, then, is one of intensely focused *lapses*, of scaleless intervals and stammering repetition of forms. Object and subject are continually distracted and superimposed by a flow of perceptual habits in a situation that has the appearance of certainty—precision, clarity, differentiation—but is only capable of provisional confirmation. Clarity, then, can introduce uncertainty, precision may be a mode of deception, and differentiation may encourage confusion. When such a situation occurs we are at the familiar threshold of randomness disguised as order.

There is a clear relation here to the space of modernist painting. Initially, modernist space was clarified so that the ambiguities of the picture plane (between two and three dimensions) were made precise. Then relational systems (along with the figure-ground relationship) were expelled. Finally, conceptual systems (aesthetic, social, historical) were invented to "read" flat, nonillusory surfaces. Similarly, the history of microscopy has concerned itself with expelling ambiguities and delusions and erecting systems by which what is seen can be read. Both developments involve the expulsion of various idealisms that donate an a priori structure to what is seen. This is not the recovery of the "innocent" eye. It is, in fact, its opposite. The eye, instructed by hosts of conceptual systems, becomes intensely sophisticated and multiple. Thoreau's doubleness is tripled, quadrupled, and more.

III

The histories of modernist art and the microscopic field are, then, entangled in the illusion of order, the belief that certain forms had attached to them a privileged status. Both inherit this belief from idealist philosophy. An idea of order is projected that when applied or fitted to certain realities and data is recovered in terms of "meaning." Meaning is lost in something, and then found again. The tautology inherent in this illusion of order haunted both modernist art and biology. It is fashionable now to look upon these traces of idealism in art with indulgent superiority. Belief makes for immobile systems, and we have developed almost a horror of such paralysis. Yet one of the phenomena of modernism was the productive nature of systems that eventually proved false. One thinks particularly of Mondrian, building an armature of ideas that his art eventually dissolves. Ever since Impressionism, systems and art have conducted a highly equivocal dialogue. But applied to the microscopic, illusions of order have proved incredibly profitable.

These ideas of order are Platonic—the spiral, the circle, the sphere, the cube. They are reducible to mathematics and thus are rational, however attended by Romantic rhetoric. Their equivalents in terms of process are repetition and symmetry. There is an emphasis on the cell, the unit, the module. Crystallography provides the inorganic prototype. As Ritterbush points out in his invaluable *The Art of Organic Forms* (Smithsonian Institution Press, 1968), the understanding of crystal structure in terms of mathematics (that is, logic) confirmed a schism in matter—between the organic and the inorganic. The organic was considered irreducible to mathematical order and thus mysterious. Life could be seen as an inexplicable vitalism on the one hand, and on the other a delusion within the inorganic, or, as it has been called, a fever in matter. Licensed by the presence of mystery, the Romantic imagination invaded the organic, which then provided Romanticism's fundamental metaphor.

Order, then, could take three forms. The idealistic blueprint derived from Platonism; the mechanist idea based on a cause and effect determinist view; and the organic, where the attempt to reduce growth and change to intelligibility has resulted in a huge literature, of which Lancelot White's *On Growth and Form* is a typical example. The organic, in its endless correspondences and relationships, becomes proto-mathematical and so appropriates some of the energies of Platonism. This complicated subject is not clarified by the fact that each of these three modes sometimes borrows from the other to describe itself. And some categories of order (for example, symmetry) tend to override the distinction between the organic and the inorganic.

The organic metaphor has been dominant in most discussions of similar images from art and science. In modernism, the organic metaphor was deeply pervasive. Process and change have been habitually applied to the development of modernism itself, to the phases of the individual artist, and to individual works of art. The creative process has been explicated in terms of biology. Predictably, German artists (those closest to idealist Romanticism) have attempted to build systems based on organic correspondence. Klee and Kandinsky are the leading examples. The Surrealist enterprise is the final apotheosis of the organic. And the biomorphic form it donated to art has been durable.

So much banal wonder and written nonsense have been provoked by the correspondences between images of art and science that it is an area where a babble of clichés holds sway. This may disguise the problem but does not remove it. There is an amusing form of interdisciplinary social climbing inherent in the comparison: science seeks an aesthetic cachet; art seeks from science the authority of which science deprived it. As Wylie Sypher makes clear, art, while opposing science, has appropriated much of its language and methods. Such sibling rivalry, however, is subsumed in the full implications of the organic metaphor.

"Organicism," wrote Stephen Pepper, "is the world hypothesis that stresses the internal relatedness or coherence of things. It is impressed with the manner in which observations at first unconnected turn out to be closely related, and with the fact that as knowledge progresses it becomes more systematized." Fundamental to organicism is the *unity* of all phenomena. All matter conducts itself according to fundamental rhythms that transcend scale. The macrocosm and the microcosm mirror each other and man mirrors both. Process is inseparable from this concept. In this view, the universe is a great pulsating beat reflected, who knows how, in the beat of the pulse at the wrist, and so on. Organicism rejects parts—it sees wholes adding up to other wholes—eventually reaching a perfect harmonious Whole. This rhetoric has been so convincing as to have an imperial power.

In this context, the correspondence between images of art and science meets no difficulty. The thinking goes somewhat like this: what is revealed by science in the process of explicating laws is revealed by the artist through an intuition of these laws. Cognitive and conative faculties are in perfect harmony and each confirms the other by different routes to Truth. Here there is a strong idealism, an old-fashioned idea of science, and the belief in what one might call the fallacy of correspondences (of which more in a moment). The idea of the artist's intuition requires further thought, for it is closely connected to Surrealism.

Surrealism ascribed to the subconscious certain functions it deemed necessary. The subconscious defines the uniqueness of each individual while at the same time declaring that all partake of the same rhythms and processes, reflecting those of the universe at large, which can only be touched by relinquishing the conscious, by automatism. In the subconscious all are brothers. All aspire to the same mythology, a quasi-religious concept somewhat equivalent to the communion of souls.

This is the theory and it is at this point somewhat ramshackle. Disbelieving in large idealisms, we can no longer accept it. But it is difficult to displace, not least because it can be replaced with very little that is convincing. The fundamental fallacy should be exposed—the implication that similarities are derived from similar processes. Analogies between images of art and science may be an accidental intersection. The problem awaits further information on man as a perceiving and concept-forming creature—an answer that safely defers it to the future.

Is it too obvious to suggest that artists, who are notoriously intelligent and searching, might, at the beginnings of Surrealism, have acquainted themselves with the results of a century of microscopic morphology? Deeply immersed in organicism, it seems unlikely that artists would not be aware of its forms. The popular science books of the late nineteenth century, when such artists were young, abounded in such illustrations. An examination of artists' journals and libraries from this point of view is indicated.

Another, perhaps stronger argument arises from the nature of art itself. European art abounds in organic contours. When suspended in the abstract, it may be true that the biomorphic form, instead of being a remarkable invention, is in fact inevitable. Not because of cosmic magnitudes, but simply because of the problems of inventing form on a surface. Certain categories and families of shape continually appear. Instead of emphasizing the multiplicity of such forms, it is possible that—like comedians' jokes being based on four or five motifs—there is a limited availability of such forms. And that invention lies in the hybridization to which they can be subjected. Similarities of color can be dismissed by pointing out that staining a microscopic section, eliciting color through the use of polarized light, and the artist's selection of color are all arbitrary—or determined. The fact of such correspondence may not be surprising. It is the habit of drawing synthetic conclusions that is.

Finally, to invoke the fallacy of correspondences. It frequently happens that the work of an artist calls forth its microscopic echo. That is, the artist has provided us with the image to identify its scientific doppelganger. This is particularly true in work that is based on all-over patterns or repetitions. Here coincidence cannot but lessen, by implication, the artist's achievement. It becomes too inviting to dismiss the artist as a literalist of microscopic nature—a Corot or Cézanne of the cellular. This is a profound misunderstanding of the nature of art and of the art of nature. It raises the question of mimesis (realism) in abstract art—doubtless a discomforting consolation to whatever enemies it still may have. The reduction of abstract art to literal genre—the logical conclusion of this assumption—raises an amusing spectre, not the least part of which is the reformation of education, since art would be brought into the area of the explicable and thus the teachable. Instead of confirming each other, such images subvert both science and art by involving them in a cat's cradle of misunderstandings. The beauty of the pictures in this book is self-referential, not relational. As such the images do not need "art" to be seen with, although art has surrounded them with very troubling echoes.

BRIAN O' DOHERTY

MICRO-ART

ART IMAGES IN A HIDDEN WORLD

INTRODUCTION

The intent of this book is to display the exciting visual images that lie beneath the threshold of normal perception. The book is therefore not essentially scientific or technological; it is not intended to prepare or instruct or train the reader. It is, rather, an *art*book—designed chiefly for the reader's enjoyment.

Photomicrography, the technique for photographing microscopic subjects, is the means by which the pictures in this book have been created. An understanding of the microscope's functions, history, and capabilities will be useful in appreciating the images that follow on these pages.

The microscope plays a greater role in our daily lives than most of us recognize. As an extension of the human eye that permits us to peer into the inner structure of matter, it has enriched the fields of biology, zoology, chemistry, and physics. It has thus become a basic tool in scientific research. An indispensable diagnostic auxiliary in medicine and bacteriology, the microscope helps to preserve our health. It examines the clothes we wear, the food we consume, the water we drink, the air we breathe, and the materials out of which we design our dwellings, vehicles, appliances, and machines. It thus enhances our technology and industrial expertise. Soil surveying, metallurgy, photoengraving, criminology, forensic medicine—there is scarcely a discipline or specialty that is not in some way indebted to this unique instrument.

THE MICROSCOPE

The word "microscope" comes from two Greek words: *mikros* meaning "little" and *skopein* meaning "to look at." A microscope is an instrument therefore that serves as a means of looking at little things. Man, with his penchant for exploring his immediate world, has probably always sought ways of examining the minute structures around him. It is likely that he chanced upon the first lens system when a single drop of water, by virtue of the convexity of its upper surface, magnified a tiny object beneath it. Among the ruins of the ancient palace of Nimrud, Sir Austen Henry Layard discovered a convex lens of rock crystal. Seneca, in the first century, described the marvels revealed through hollow spheres of glass containing water; while Pliny recorded observations of objects studied under water-filled globes. Without some kind of magnifier, it would have been impossible for the ancients to practice their exceptional skill at gem-cutting.

About 1590, two Dutch spectacle-makers, Hans and Zacharias Janssen, constructed what is considered to be the first microscope. It consisted of a tube with a lens on either end—one near the object (*objective*) and the other close to the observing eye (*eyepiece*). This design provided the basic concept of the modern microscope. Descartes in 1637 described how a concave mirror in conjunction with a lens could illuminate an object to great magnitude. But the first real innovator of the practical microscope was a Dutch naturalist, Anthony van Leeuwenhoek, who, about the end of the seventeenth century, ground excellent single lenses of short focus that enabled him to study intimately the tissues of animals and man and to arrive at some interesting hypotheses. He made pioneer discoveries in regard to blood capillaries, and he rendered the first accurate description of human red blood corpuscles. Through the use of his early microscope he was able to dispel a few myths regarding the spontaneous generation of some lower animals that were reputed to be "bred from corruption." For instance, he proved that the lowly flea underwent an exciting metamorphosis from egg to adult, countering the notion that "this minute and despised creature" was produced from sand, dust, the dung of pigeons, and urine. By the same token he rebuked the ideas of the "scientific"

Anton van Leeuwenhoek, 1632–1723

euwenhoek's flea microscope, late 17th century

men of the times that shellfish were created from mud and that eels emerged out of dew.

Even before the period that Leeuwenhoek was perfecting his *simple* microscope, experiments were going on to magnify an observed image by interposing a second lens system between the lower, or objective, lens and the eye. Among the innovators were Zacharias Janssen and Galileo. The resulting *compound* microscope, while theoretically feasible, did not prove practical in use. Such eminent experimenters as Sir Isaac Newton and later Giovanni Amici worked diligently on the construction of the compound microscope, but it was not until the middle of the nineteenth century that true progress was registered. Arranging several achromatic lenses superimposed upon each other, Charles Chevalier and Joseph Jackson Lister succeeded in reducing the distortions produced by those lens aberrations found in simple (uncorrected) lenses. Around 1846, Carl Zeiss began manufacturing compound microscopes at Jena, Germany, and twenty years later Ernst Leitz developed his factory at Wetzlar, Germany. Two important manufacturers entered the field in the United States, Charles Spencer and Edward Bausch.

In 1886, Ernst Abbé invented the supersharp *apochromatic* objective, a lens completely free of flaws (that is, chromatic and spherical aberrations). Thereafter, a host of new discoveries—compensating eyepieces, sophisticated illuminating systems, vertical illuminators, phase-contrast objectives and condensers, interference and polarizing apparatus—have revolutionized microscopy and have enabled us to examine intimately the composition of organic and inorganic matter.

The *electron* microscope, a twentieth-century invention giving magnifications many times greater than the compound microscope, has opened up the exploration of infinitesimal objects previously inaccessible to the human eye. With new methods for the preparation of biological materials, it has allowed glimpses into the structural details of cells down to sizes of large molecules; instead of an undifferentiated ground substance in cytoplasm (the body of the cell) as seen with the compound microscope, the electron microscope reveals systems of filaments, rods, and other structures, giving us a better basis for understanding the very nature of the cell. The field of virology, that is, the study of protein particles that invade cells, has blossomed as a result of the electron microscope. With each passing year, improvements on this instrument, leading to greater efficiency of operation, are increasing substantially the range of our vision.

left: *Hand microscope, early 18th century*

Microscope, c. 1740

bottom: *Microscope, 1850 (Carl Zeiss, Inc.)*

Photomacrography and photomicrography

If we are to record images in the world beyond the range of normal vision, we need specialized equipment. Basically this consists of an optical system, a source of light, and accessory equipment to photograph the microscopic objects in our field of vision.

Where magnifications of up to fifty diameters are desired, a simple microscope is employed. As described earlier, this microscope consists of a single lens poised over the object to be observed, which, should we wish to employ photography, is attached to a camera by an extension tube. The product of this effort is called a *photomacrograph*, the process *photomacrography*. The simple microscope covers a wider field than a compound microscope, enabling us to observe entire objects such as small insects, seeds, coins, and so on. With a compound microscope, on the other hand, the view of a cross section of a plant stem, for instance, would be circumscribed to a tiny part of the stem section, permitting us to see this in great detail but eliminating all peripheral areas.

In photomacrography, intimate features are sacrificed in favor of the greater field of vision, desirable for the study of larger specimens. One advantage of using a simple microscope is that it permits us to bring into focus relatively thick materials, which at greater magnifications would be blurry. Should we desire more detail than is afforded

Microscope, c. 1878 (Carl Zeiss, Inc.)

right: Electron microscope (Carl Zeiss, Inc.)

bottom: Microscope, c. 1920 (Carl Zeiss, Inc.)

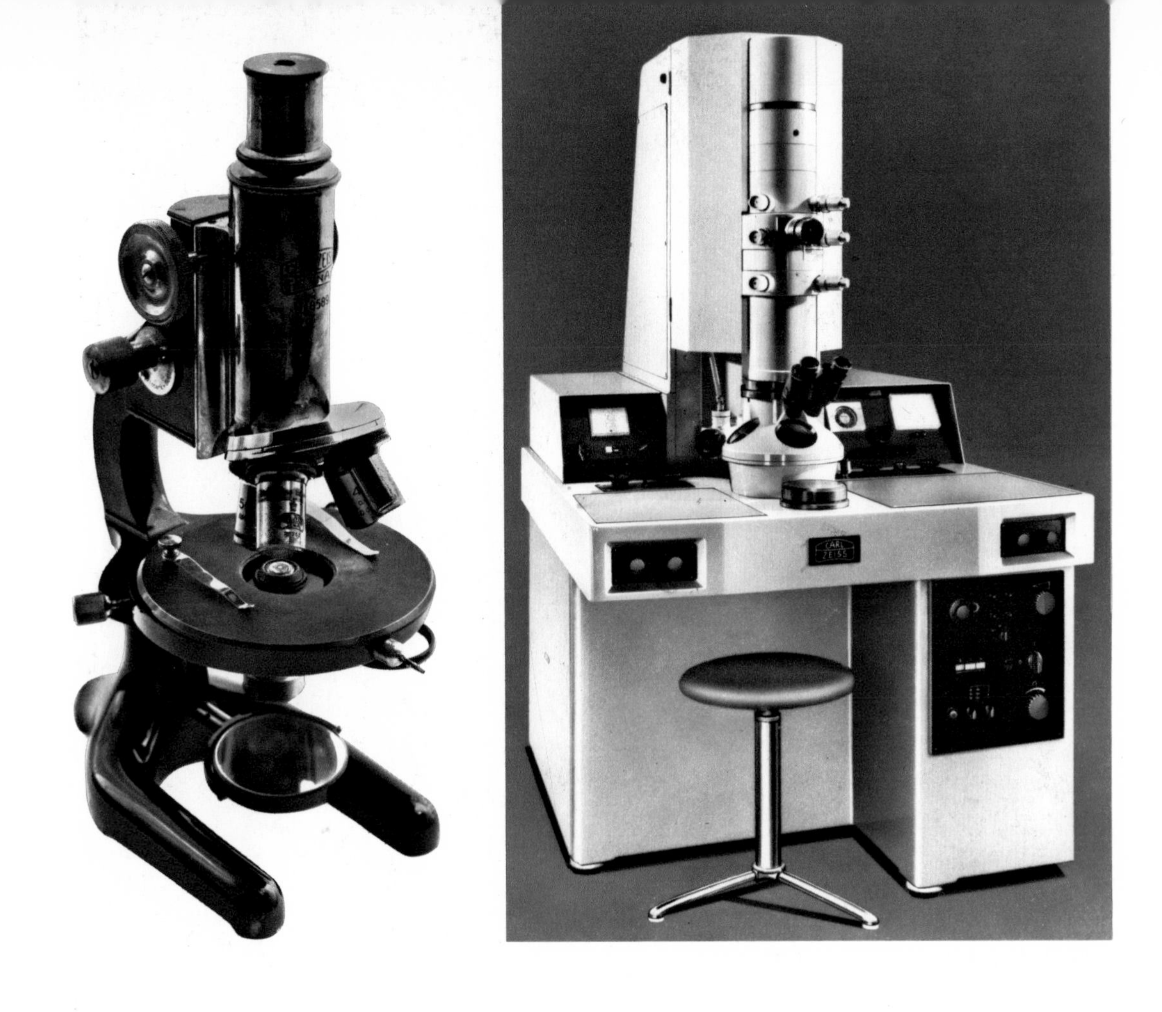

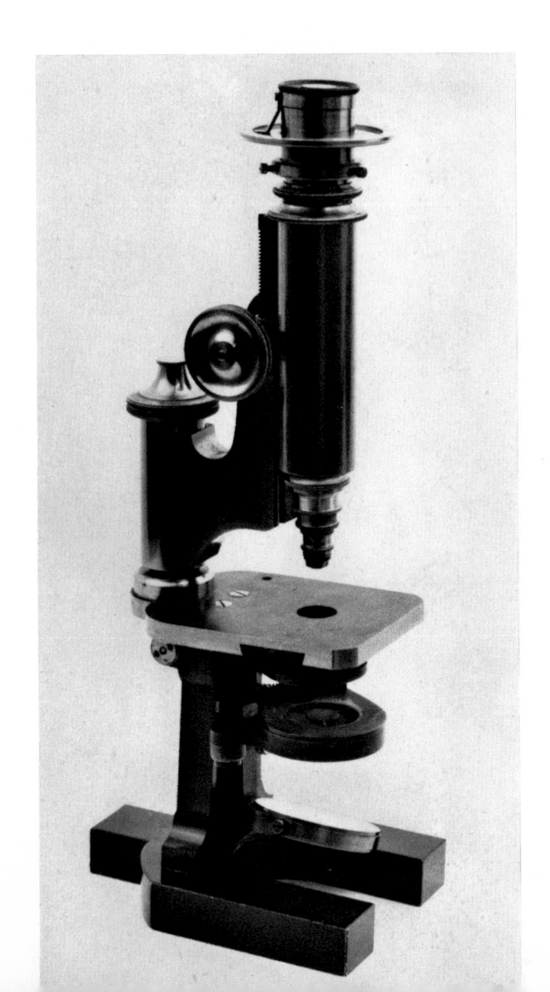

through photomacrography, we must employ the higher magnification of the compound microscope. The basic optical system of this instrument consists of two lens arrangements: the objective and the eyepiece. When a camera is attached to the eyepiece, and we photograph the specimen, the process is called photomicrography.

Lenses

Since in photomicrography we are working with visual minutiae, we require lenses in our objective and eyepiece that are highly corrected for the tiniest flaws. We also have to be able to *resolve* the highly magnified image—which means we would require a lens that can separate adjacent particles and bring them into sharp focus over the entire observational area. The most commonly used objective lenses in microscopes are achromatic. Apochromatic lenses are best suited for high-power photomicrography. Ranging midway in quality between the achromatic and apochromatic lenses are the *fluorite* lenses. Recently, some highly improved lenses have been introduced that, in addition to rectifying lens aberrations, produce a uniformly sharp field; they are *plano-*

achromatic lenses and *plano-apochromatic* lenses, the latter being of the very finest quality.

Lenses are graded by a designation known as a *numerical aperture* marked on the outside of each objective lens, which signifies its capacity to differentiate detail (resolving power). This gives us a measure of its ability to allow for an optimum degree of picture enlargement in photomicrography. The larger the numerical aperture of a lens, the greater the magnification we can achieve. If we enlarge beyond the potentialities of the designated numerical aperture of a lens, we will end up with an unclear print, a condition often referred to as "empty magnification."

Like camera lenses, objective lenses are available with different focal lengths that provide for different degrees of magnification. For instance, objectives marked 4x, 10x, 16x, 25x, 63x and 100x will enlarge the viewed object precisely that many times. Where we wish to secure the highest degree of magnification and resolution, using objectives of high numerical aperture, we employ lenses especially constructed to be immersed in cedarwood oil (or synthetic oils). An oil-immersed objective reduces the dispersion of light—unavoidable in the air gap between dry objectives and the slide—and helps to sharpen the image.

The second lens system of the compound microscope is contained in the eyepiece, which is usually matched to the finer objectives and compensates for residual errors in them. The eyepiece augments the image of the objective by the magnification factor marked on it. Thus the 8-power and 10-power eyepieces utilized in the making of the photographs in this book boosted the 4x objective to 32 and 40 powers, respectively; the 10x objective to 80 and 100 powers.

Objectives and eyepieces are fastened to a specially designed body of the microscope that allows for a variety of attachments. The most expensive bodies permit rapid changes of the various components of the microscope (eyepieces, objective lenses, and so on). They usually have an efficient, built-in lighting unit easily centered by a special lens that focuses light onto the specimen; and they provide for accessory lighting units used to achieve unusual effects. The less expensive bodies are quite serviceable, although they require more time to allow for essential adjustments, and they are not as flexible in adapting the attachments.

Light sources

Of vital importance in microscopy, and especially photomicrography, is the illumination source that brings out the details in a specimen. The basic unit is a beam of light that penetrates the slide from below. This form of illumination *(bright-field)* is used

Photomicroscope (Carl Zeiss, In

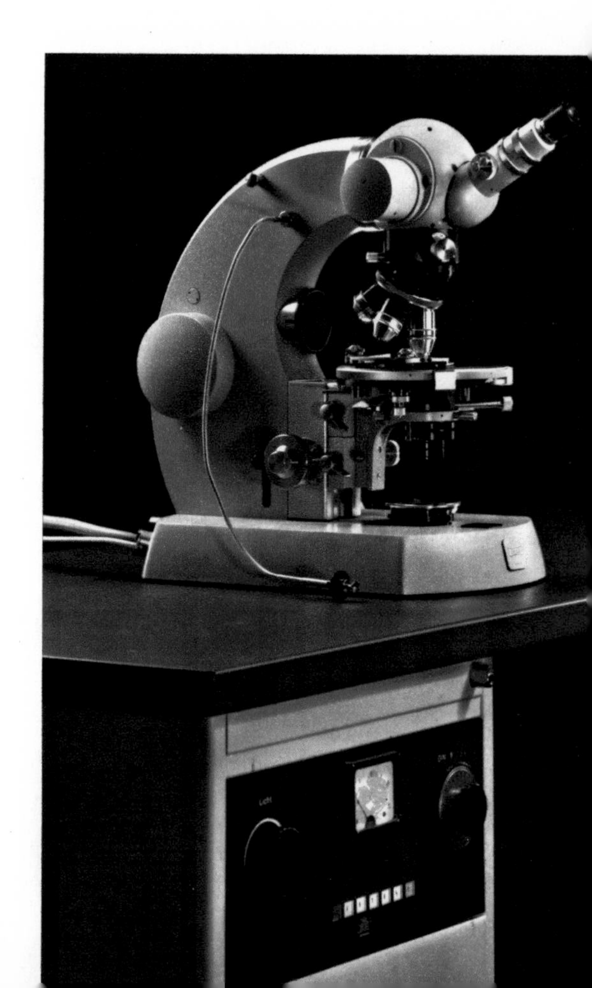

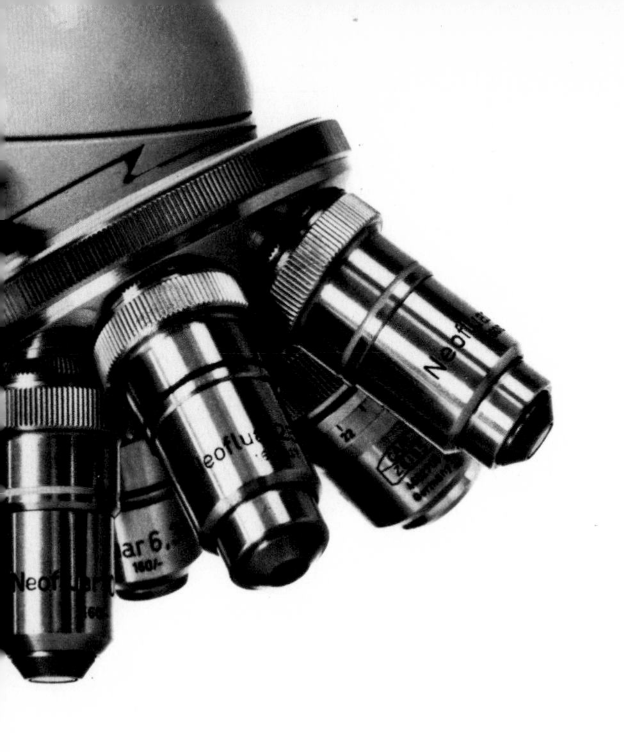

'osepiece with neofluors (Carl Zeiss, Inc.)

for transparent objects, which in fact constitute the bulk of the materials reproduced in this book.

In photographing the minute objects under scrutiny, relatively intense lighting is required. This light must have certain characteristics, including a compatibility with the film that is used. The basic source of illumination is supplied by a small, incandescent tungsten-filament light bulb (either 6 volt/15 watt or 12 volt/60 watt) contained in a separate lamp housing and placed alongside the microscope or built into its base. The light is reflected off a tilted mirror, or projected through the built-in light system. A *diaphragm* controls the size of the light disc, and a *field* condenser concentrates it.

Bright-field illumination, as mentioned above, is most commonly used where light is required for the viewing of slides (*transmitted* light). Here the beam of light is passed through the *substage* condenser, which narrows the light down to a solid intense cone; this process is known as *critical* illumination and involves the correct positioning of the condenser in relation to the specimen and the proper adjustment of the opening of the condenser (*aperture diaphragm*). In short, the substage condenser is moved up or down until its diaphragm opening is completely filled with even lighting, which is in turn focused onto the plane of the specimen. The diaphragm is then adjusted for optimum contrast and depth of field, a technique (*Kohler* illumination) which gives the brightest and sharpest light attainable.

Light sources other than incandescent tungsten-filament lamps are less commonly employed. Where a very intense light is needed, as in metallography, a *carbon arc* lamp is useful. A *xenon arc* lamp produces a very strong light of daylight quality, useful for specialized lighting effects (as in phase-contrast and interference microscopy, about which more will be said later). A *mercury-vapor* lamp emits ultraviolet rays of short wavelength that yield sharper images than can be obtained with incandescent light. The *zirconium arc* lamp is intense and has certain advantages in being quite small in size. Finally, *electronic flash* and *wire-filled* flashbulbs may be used for photographing moving objects.

Some structures do not show up clearly with conventional bright-field illumination. Displacing the light slightly so that it illuminates the specimen from the side, or obliquely, sometimes serves to produce three-dimensional effects that are of pictorial interest. Colorless unstained objects tend to blend themselves with the background so that they can hardly be seen with bright-field lighting. In these cases, oblique lighting may help bring them out into visible perspective. More effective however, in such cases, is *dark-field* illumination. Here the central rays of light are blocked off by interposing between the condenser and slide an opaque circle that is small enough to permit peripheral rays to glance off the objects, thus lighting them up in a sculptured white against a black background. Many unstained bacteria, crystals, and colloids that ordinarily are unseen reveal themselves in this way.

An important contribution to lighting in microscopy was made in 1935 by Frits Zernike; known as *phase-contrast,* it differentiates translucent objects from a background that would otherwise render them invisible. Special condensers and objectives are required to reveal the slightly varying densities of objects and their backgrounds. Unlike dark-field illumination, phase-contrast brings out the inner details of structures. It is therefore widely used in the study of living cells, particularly in the field of tissue culture.

Another recent advance in microscopic lighting—*interference-contrast*—reveals images free of halos that are sometimes created by the phase-contrast technique. Interesting color effects may be obtained through interference lighting. The equipment is somewhat elaborate but definitely worthwhile when photomicrography is employed for artistic effects.

Interesting colors may also be obtained by utilizing a mercury-vapor lamp as the lighting source. The ultraviolet and blue rays emitted cause certain substances to change color and glow (fluoresce). For instance, chlorophyll will change from green to deep red, and certain stains will light up brilliantly. Employing dark-field illumination together with fluorescence microscopy produces a startling picture of a luminous object against a dark background.

Where opaque objects must be photographed by transmitted light, infrared rays will penetrate a specimen that otherwise would be impervious to ordinary light rays. Specimens such as whole insects, dark crystals, textiles, and paper are more suited to infrared than to conventional lighting. Special infrared filters and films are required for this purpose.

Perhaps the most common form of special-effects illumination is *polarized* light, easily obtained through the use of polarizing filters—one placed above and one below the specimen. Striking effects may be obtained through polarization. Crystals, rock sections, hairs, various tissues, and certain transparent objects all sparkle in color against a dark background; if, in addition, a quartz filter is placed in the beam of light, the background will stand out in a contrasting color.

Opaque objects, such as metals or thick-textured materials have special illumination requirements. Here a *vertical* illuminator is necessary, the light being reflected from above through tilted glass or a prism. This is known as *incident* lighting.

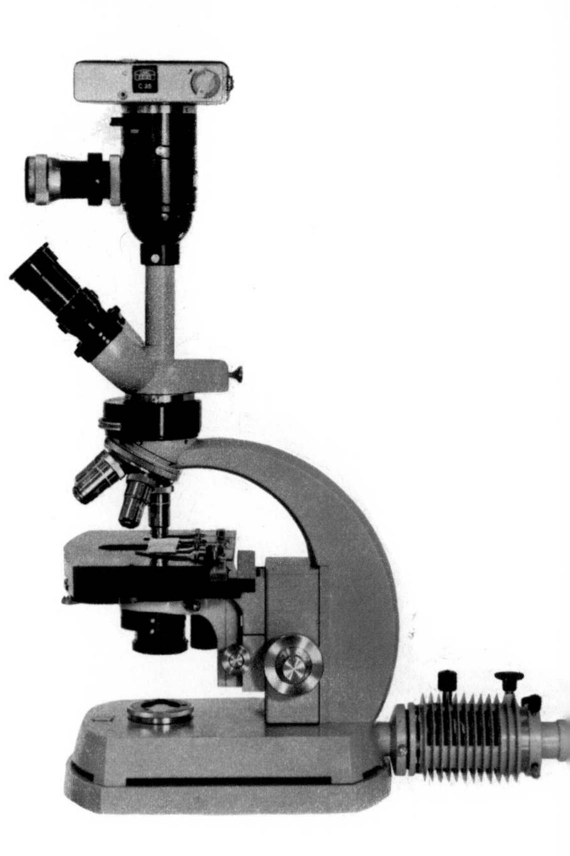

Microscope with 35mm attachment camer

(Carl Zeiss, Inc

The photomicrograph

The easiest part of photomicrography is taking the picture. Practically any camera may be adapted for this purpose. The simplest, though least satisfactory, setup is the employment of a camera with an integral lens, the front surface of which is placed as near to the eyepoint of the microscope as possible. The vertical bar of a copying or enlarging stand holds the camera in place. Since the image of a focused microscope is considered to be at infinity, the camera distance-setting should also be at infinity. For certain 35-millimeter cameras, microscopic adapters are available with fixed lenses that obviate the need for a copying stand.

A much more workable setup uses the *eyepiece* camera—specially designed for photomicrography. The best camera apparatus is one with a movable bellows, a groundglass screen for focusing, and provisions for removal of the front lens. Fastened to a copying stand, and equipped with a microscope adapter in place of the lens, the camera is lowered onto the eyepiece, which in turn is fitted with a complementary adapter. This arrangement prevents external light from entering the camera. The cameras employed for making the pictures in this book are of this type.

Slides

Apart from metals and other opaque materials, most objects are studied under a microscope on a thin glass slip. Usually, a specimen is mounted with a resin that hardens and then is covered with an extremely thin glass *coverslip*.

Where the object is mounted in its entirety, it is called a *wholemount*. Most materials, however, are too large for wholemounts. They must instead be cut into small pieces, hardened, dehydrated in successively stronger concentrations of alcohol, dealcoholized, and impregnated in molten paraffin. The hardened paraffin block is then mounted in a slicing tool (*microtome*), and an extremely thin shaving is sliced off. This shaving is placed on a slide, and the paraffin is removed by a solvent. The specimen is then stained and mounted. Since this process is a laborious one, it is fortunate that many excellent, permanently prepared slides may be purchased from special biological supply houses.

ᐩicroscope with fluorescence illuminator

ᑐarl Zeiss, Inc.)

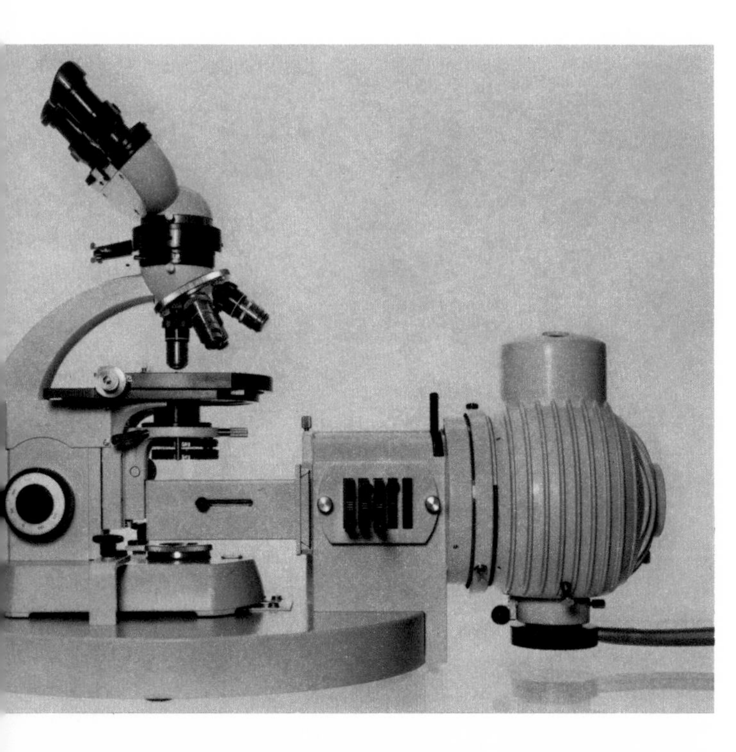

Staining

A good deal of what we know about the structure and function of cells and tissues is through the use of selected stains. Since different portions of a given cell may have varying chemical compositions, they will react with certain stains more readily than with others. Thus parts of a slide will pick up some stains and other parts reject them. This permits the differentiation of many inner structures that otherwise would escape notice. Staining is consequently an indispensable technique in microscopy.

The process of staining is complex. A common staining practice is to immerse the object in carmine or hematoxylin solution, then in acidified alcohol to remove the dye from some portions. The specimen is then dehydrated in successively more concentrated alcohol solutions, after which it is dealcoholized in xylol. This allows the resin to penetrate the object.

Another staining method starts by smearing the specimen itself in a thin layer on the slide. This process is obviously suitable for soft or liquid materials, such as bacteriological cultures or blood. When the smear has dried, it is stained; then, after a minute or so, the slide is rinsed with water or buffer solution. The smear, now stained, after drying may be examined as it is, or it may be permanently mounted with a resin and covered with a coverslip. In preparing bacterial specimens, the material is usually transferred to the slide with a freshly flamed and cooled small wire loop and allowed to dry. It is then heat-fixed in the flame of a spirit lamp, covered with a stain, washed with a jet of water to remove the excess stain, and again permitted to dry. If the slide is to be kept permanently, it is topped with a drop of resin and a coverslip.

Although there are many more staining techniques available to the micrographer than those discussed in this introduction—techniques usually applied to particular materials for specially desired effects—further description of staining procedure would not be useful here.

It is well to note that the staining procedures followed for all the illustrations in this book are in complete accordance with standard laboratory practices. That is to say, no staining was done to heighten the microscopic images any further than is normally done for scientific examination.

Dog flea, actual size

NATURE, ART, AND THE MIND

If we examine a drop of pond water under a microscope, we will immediately be confronted with a startling panorama of beauty and movement. We see green-bellied *Euglena* tumbling over each other or circling around finely sculptured diatoms. A slipper-shaped *Paramecium*, with cilia fanning particles into its oral groove, dines on yeast and bacteria. Strands of *Spirogyra* in luminous coils seem to provide an ambush for the many infinitesimal predators in this strange microjungle. Suddenly a *Didinium*

Macrophotograph, dog flea, 32x

right: Macrophotograph, dog flea, 50x

darts out to seize and swallow the *Paramecium*. A sluggish *Amoeba* thrusts out a lazy pseudopod from behind a shell, its precarious outpost in a Lilliputian land of parasitic marauders. And, amidst this spectacle of life and death in a fragile universe of little things, we are startled by the interlacings of textures and shapes. Inanimate motes are converted into mountains of beauty. Tissues of living cells are strung together exquisitely like bands of leisurely filigree. As we shift the slide around, a succession of vignettes envelops us in an environment of visual excitement. Viewing these natural masterpieces of "nonobjective art," it is gratifying to observe the subtleties of color, the gradations of texture, the magnificence of design, and the brilliant shadings of light that have taken nature hundreds of millions of years to fabricate.

People almost instinctively respond with pleasure to certain uncontrived forms. Flowers in bud or full bloom, bodies of water surrounded by ice-capped mountains, a full moon suspended in the sky, trees silhouetted against a setting sun—these and other images arouse in us almost universal acceptance. Seemingly, it is the prerogative of nature to adorn her estate with artistic trappings and juxtapositions to which man assigns esthetic value.

Nature in her magnificence spreads herself over a prodigious area, from the systematized arrangement of ions to the orderly configurations of stars and planets. Our microscope reveals the most intimate tableau of nature's handiwork, the embellishments of inorganic and organic matter fashioned by a cosmic design of adaptation. It is a hollow challenge to expect man to approach this versatility in his art forms. Yet it is wondrous to observe how a skilled abstractionist can stir up feelings akin to those released by nature's compositions.

There are, of course, those artists whose works objectively communicate some approach to the visual analysis of nature—for example, Dürer, Alberti, Masaccio, Leonardo, and Pollaiuolo. But what is astonishing is that the productions of certain abstract artists resemble in a startling way the configurations that one sees through a microscope. Indeed, were we to place photographs of microscopic objects next to the works of some well-known painters, we might be tempted to say that they were mastered by the same hand. We may be at a loss to explain what seems to be more than a mere coincidence. Have these painters for their inspiration drawn from nature's reservoir by deliberately peering through a microscope? Have nature's designs registered themselves in the brain through some other experiential means? Does man carry within his memory residues of his evolutionary past, the archetypes of his own vestigial forms, with primitive cellular images imbedded permanently in his mind?

We may speculate that every person—the artist in particular—subconsciously records images from real experience that may later come to consciousness as designs or patterns which he believes to be internally inspired. By the same token, what we see under a microscope may in fact exist in some related form visible to the naked eye. Certainly, we should not dismiss as fanciful the arguments of scientists, philosophers, and even poets who have offered certain esoteric explanations of these phenomena. Perhaps someday their theories will be experimentally validated. It was Dickinson W. Richards who wrote in the *Transactions of the Association of American Physicians* (75:1, 1962): "We are forgetful of the curious and wayward dialectic of science, whereby a well-constructed theory, even if it is wrong, can bring a signal advance."

Why do the representations of some artists so often resemble microscopic structures? Several theoretical possibilities pose themselves: (1) Coded and stored inner arrangements (engrams) may enable the artist to objectify the order and composition of the universe. (2) The cells of the brain and the perceptual apparatus possess basic ingredients and patterns found in all organic and inorganic matter; these cells may under certain circumstances project images of their own nature. (3) Certain gifted persons may be capable of recording in creative expression the sequential movement, symmetry, balance, and proportion that are inherent in their perceptual minds and that also correspond to natural forms, including microscopic forms. Thus man may be responding to the same interacting processes that operate in all of creation. As Emerson expressed it in his *Nature:* "Compound it how she will, star, sand, fire, water, tree, man, it is still one stuff, and betrays the same properties."

The idea that man's brain is capable of delineating its own contents is not a new one. Ancient philosophers, including Plato, embraced it. Eastern religions, such as Brahmanism and Buddhism, promote the concept of an eternal essence that permeates all of nature, including man, thus enabling one to absorb the Absolute through meditation.

In the Yoga system, the follower who arrives at the last three stages of concentration (samyana) achieves an ecstatic mystical state during which he is said to have an awareness of past and future and a knowledge of everything that exists in the world, including basic matter. The satori experience in Zen produces what is often envisaged as a fusion of the mind with the whole of nature. In Western culture, German metaphysicians organized some stimulating speculations around the same premise. For instance, Kant speaks of a unitary thing-in-itself that permeates the real world but that mysteriously is also experienced in the human mind. Hegel, posing an Absolute substance that exists in both nature and the mind, avows that nature is the medium in which the Absolute works out its logical schemes in terms of time and space.

Poets, like Wordsworth, Shelley, and Tennyson, have referred, if somewhat obliquely, to the role of nature in the human cogitative scheme; and Thoreau has written of the unity of man with the constituents of the natural world. Some psychological schools have endorsed the notion that the brain stores visions of the past and can on occasion revive them symbolically—and conceivably in the form of a painting. The Gestalt school suggests that innate, or inherited, physiological laws actually govern the ways by which the individual perceives tangible objects in the real world. Finally, neurophysiologists have proposed many similar formulations, though in more technical terms.

These theories may be scientifically questionable when applied to the premise that abstract artists paint images based on their inherited visions of nature. Perhaps a more elemental explanation is closer to the truth. After all, the rhythms of nature do repeat themselves from molecules to mountains. Matter reflects a remarkable unity in patterns of energy conservation and release that seem to be the inherent property of both living and nonliving things. Some of the arrangements we see through a microscope in all probability have identical counterparts in macroscopic, or larger, forms. Patterns and designs thus will tend to duplicate themselves throughout nature.

Answers to our questions about the relationship between art and microscopic forms must await the expansion of our knowledge that lies in the future. In the meantime, we may be granted dispensation to enjoy the great continuity of unique forms, textures, and patterns that the microscope with its optical devices has opened up to man's perception and interpretation.

55. Hollyhock pollen (multi-stained)
950x

56. Blue-green algae: *Rivularia,* basal heterocyst
230x

57. Pine needles
160x

58, 59. Diatoms: *Pleurosigma angulatum*
1,200x Polarized light

60. Eastern hemlock *(Tsuga canadensis)*
300x

61. Bald cypress *(Taxodium distichum)*
1,200x

62. Airborne seed
190x Polarized light

63. Airborne seed
300x

64. Fungus: *Saccharomyces*
3,000x Polarized light

65. Cucumber wilt
500x

66, 67. Green algae: *Laminaria* blade with sporangia
375x Dark field

68. Desmids
750x

69. Virus
160,000x Taken with an electron microscope

70. White oak *(Quercus alba)*
300x Polarized light

71. White oak *(Quercus alba)*
300x

72. Diatom: *Stauroneis* (bacteria in background)
1,690x

73. Leaf skeleton
225x Dark field

74. Fungi: *Ascomycetes*
700x
Fruit cells *(microphaera ascocarps)*

75. Eastern hemlock *(Tsuga canadensis)*
230x

76, 77. Pine stem
320x

78. Dicotyledonous plant: *Tilia* stem
180x Dark field

79, 80. Beech *(Fagus grandifolia)*
490x

81. Sponge: *Sycon ciliatum*
1,000x Polarized light

82, 83. Bamboo *(Bambusa)*
600x Polarized light

83. Bamboo *(Bambusa)*
600x

84. Green algae: *Fucus,* male conceptacle
700x

85. Liverwort
3,000x Polarized light

86. *Volvox*
525x

87. Diatoms
750x

88. Potato scab
750x Polarized light

89. Fungus: potato scab
190x Polarized light

90, 91. Pine *(Pinus banksiona)*
1200x

92. Green radish root
300x Polarized light

93. Moss: *Mnium*
300x
Female sexual organs *(archegonia)*

94. Leaf stalk of garden radish
500x

95. Leaf stalk of garden radish
565x Dark field

96. Green algae: *Vaucheria*
200x

29. Diatom: *Pinnularia*
2,800x Interference contrast
Courtesy of Carl Zeiss, Inc., New York.

30. Diatom: *Synedra*
1,200x

31. Pollen (stained)
210x

32. Algae: *Spirogyra*
700x Polarized light

33. Diatoms: *Pleurosigma angulatum*
500x

34, 35. Outer onion skin showing crystals
900x Polarized light

36. Paper birch *(Petula papyifera)*
600x Polarized light

37. Diatom
2,000x

38, 39. Diatom interior
6,900x

40. Diatoms
top, 560x bottom, 530x

41. Blue-green algae: *Oscillatoria*
475x

42. Douglas fir *(Pseudotsuga taxifolia)*
750x

42, 43. Holly *(Ilex opaca)*
750x Polarized light

44. Marine plankton
400x Polarized light

45. Marine plankton
600x Polarized light

46. Diatom: *Synedra*
3,000x

47. Mixed pollen (stained)
130x Dark field

48. Cottonwood *(Populus deltoides)*
760x

49. Hard maple *(Acer saccharum)*
315x

50. Northern pine *(Pinus strobus)*
200x Polarized light

51. Northern pine *(Pinus strobus)*
200x Polarized light

left: 52. Diatoms: *Stephanodiscus*
1,200x

right: 52. Green algae: *Draparnaldia*
500x

53. Beech *(Fagus grandifolia)*
190x Polarized light

54. Diatom: *Trinacria*
1,200x

Keep fold-out open for captions to pictures in the following section

VEGETABLE

Keep fold-out (at left) open for captions to the following section

All living things are divided into two kinds of matter: vegetable and animal. While it is easy enough to distinguish the features that characterize the higher orders of these groupings, the lower forms—particularly the unicellular species—are quite hard to differentiate, since they stem from a common ancestor. The main distinguishing factors are a) nutrition: plants usually produce their own food from simple substances, while animals are parasitic; b) growth pattern: plant tissue displays continuous growth, with cells accumulating to expand the size of the plant, while animal tissue does not grow beyond a fixed adult size; c) cell walls: plant cells are constructed from rigid cellulose, while animal cells are soft and elastic; and d) locomotion: most plants are anchored to some solid body and cannot move of their own accord, while most animals are mobile. The photomicrographs in this section range from the tiniest virus—the link between living and nonliving matter—to specimens of the largest plants—trees—as illustrated by sections of wood.

One-celled plants

The earth is barely a pinpoint in a huge galaxy that spreads across the skies in myriads of stars. Within our telescopic sight are one hundred million such galaxies, each with its great stellar multitude. Our own star, the sun, which we share with other planets, is believed to have originated almost five billion years ago from the condensation of gases, dust, and ice swirling in space. The stored heat and energy in the sun, a consequence of this condensation, exploded thermonuclear reactions, which, as the earth was being formed, penetrated our atmosphere in rays. Propitious circumstances of temperature, moisture, atmospheric content, and energy released from heat and lightning favored the earth as a site for the formation of living systems. Experimentally, similar conditions for the formation of viable systems have been reproduced in the laboratory: by subjecting the materials of atmospheric content to high-frequency electric charges or ultraviolet rays, scientists have created amino acids, the essential building blocks of living things.

In the gigantic crucible of our atmosphere, large quantities of amino acids were believed to have been formed during the earth's first billion years and swept in masses into the oceans and onto its surface. Aggregates of organic molecules (*coacervates*) accumulated, and at some point, it is hypothesized, acquired such large mass that they began to split into two identical units, a primitive form of self-reproduction. When life actually began is difficult to say, but fossils of bacteria-like organisms have been discovered that date back three billion years. These were the original units (*autotrophs*) that survived in the ocean because they were able to produce complex chemicals from the simple ones that enveloped them.

At some point in the evolution of life a remarkable phenomenon occurred: *photosynthesis*, which enabled certain cells, because of their chlorophyll content, to exploit the energy of the sun to convert carbon dioxide and water into glucose—the fuel for living activity and the basis for manufacturing most other essential cellular ingredients. The breakdown of glucose for the energy needs of the marine cell was accompanied by the discharge of oxygen, which activated other forms of metabolism and also accumulated in the atmosphere. The atmospheric oxygen filtered out harmful radiation released by the sun, and this, in turn, allowed cells to rise from the protective shield of the ocean's waters to emerge onto land. Without the available oxygen in the atmosphere, crucial for metabolism, these cells could not have survived.

In this section we shall consider some primitive forms of plant life that, while they originated much earlier, flourished in the Cambrian period of the Paleozoic era, that is, between five and six hundred million years ago. It is possible that many of our present-day unicellular organisms are the direct descendants of these archaic progenitors and that they possess similar habits and physical structures. Undoubtedly, mutations have occurred that have produced variations in form and function. New species have therefore probably evolved through the process of natural selection.

Viruses

Scientists speculate that inorganic entities became welded together with organic matter to produce certain *organisms*. Our present-day viruses may represent this link between the nonliving and living. Virus particles, too small to be seen with the optical microscope, are visible with the electron microscope in various shapes characteristic of their species; the most common shapes are spherical, rod-like, and oval.

Bacteria

Bacteria most closely resemble organisms similar to the earliest forms of life on earth. Three major groupings occur: the rod-shaped *bacilli*, the spherical *cocci*, and the spiral-shaped *spirilli*. Bacteria are hardy cells, which can survive the most difficult of environmental conditions. They multiply rapidly in a favorable milieu. More than three thousand species of bacteria exist, and by sheer aggregate weight they exceed all other living forms. Many species are parasitic, but only some of these are disease-producing. Most bacteria are harmless; many indeed are essential for sustaining life processes, such as food digestion, in both animals and man. Soil-dwelling bacteria are valuable in producing essential nitrogen, which improves soil fertility. Saprophytic bacteria live off dead organic material, breaking it up into substances that enrich the soil.

Fungi and algae

Two large groups of unicellular plants, the fungi and algae, are classified under the division *Thallophyta*. Fungi lack chlorophyll, hence cannot exist independently. They derive nutrition parasitically from living matter or saprophytically from dead organic matter. Yeast, composed of a group of single cells, is a commonly used fungus—important for brewing, bread baking, and the manufacture of liquors and wines.

The algae are, from the standpoint of the photomicrographer, more interesting. The ingenious shapes of these creatures are fascinating. A particular group of algae, *phy-*

toflagellates, possesses properties of both plants and animals. Among these can be found the common *Euglena,* which is shaped like a submarine. It accumulates in small bodies of water, and, since it constitutes the bulk of algae scum, is considered a nuisance by owners of swimming pools and ponds. It is a highly active plant/animal propelled by a flagellum (a whip-like tail) toward any available source of light. Another common form of algae is *Peranema,* which looks identical to *Euglena* but has no green content (*chloroplasts*).

The ingenious shapes and forms that algae assume are many and are distinctive for each genus. *Phacus* is cone-shaped; *Closterium* is formed like a crescent; *Chlamydomonas* looks like a swimming watermelon; *Scenedesmus* is a horn-shaped collection of oval or pointed cells; *Gonium* is a cluster of four to sixteen cells arranged in a flat or slightly curved disk. *Pandorina* is a sphere of sixteen tightly packed cells with the flagella of each cell pointed outward. *Volvox* is a hollow globe, the outer surface of which is composed of from five hundred to fifty thousand individual cells held together by fine protoplasmic strands making a hexagonal pattern; the familiar *dinoflagellate* has a groove encircling its body like a girdle and revolves like a top.

Among the useful one-celled algae are the diatoms, which Darwin described as the most beautiful creatures on earth. There are over ten thousand species of diatoms, each individual being composed of two symmetrical valves encased in a boxlike enclosure and presenting geometric shapes of great diversity. Both fossil and living varieties have lovely, patterned shells varying in color, size, and organization of valves. Some sparkle like gems whether viewed with direct or polarized light. Fields of diatoms under dark-field illumination glow like galaxies of stars. Diatoms constitute the bulk of the plankton eaten by protozoa, worms, mollusks, fish, whales, and other denizens of the sea. They are therefore of great ecological importance. Approximately sixty to seventy million years ago, in the Tertian period of the Cenozoic era, they underwent a spectacular multiplication. Indeed, their shells have accumulated to such extents that they have created entire islands.

Many-celled plants

The single living cell—whether plant or animal—is a microscopic vessel that functions as an autonomous unit. It can eat, excrete, reproduce, attack, and defend itself against injury. In its fight for survival it can shift the mode of its functioning and, through mutations, even change its form. Thus it may band together with other similar cells as a member of a colony, each cellular unit operating independently. Or, it may alter its operations to become a single building block in a colony, forming tissues that within the larger organism have specialized, or limited, functions, such as feeding, excreting, locomoting, reproducing, or attacking.

Cells are intricate manufacturing units, capable of complicated exchanges of energy intake and output for purposes of metabolism and other life functions. An outer membrane of tough tissue encases the cell's contents yet permits the movement of substances into and out of the cell by means of *osmosis*. An inside cellular membrane, so thin that it is almost invisible, contains the *cytoplasm*, a jelly-like mass comprised of fat, crystals, granules, yolk spheres, and watery vacuoles.

In the *embryo* of the multicellular organism, rapidly developing individual cells arrange themselves in groupings and assume structures and functions assigned them by genetic determinants. The *nucleus* contains the hereditary determinants that dictate the behavior and destiny of the cell, while the cytoplasm governs processes such as respiration, differentiation, and secretion. These two divisions of the cell are mutually interdependent.

The microscope has permitted us to solve some of the mysteries of nuclear structure and function. It has shown the nucleus to be enclosed in a *nuclear envelope* perforated with pores through which large molecules enter the cytoplasm. Inside the nucleus is a spherical body, the *nucleolus*, which is responsible for the manufacture of the ribosomes, granules that produce proteins.

With our microscope we can examine the evolution of the organism from fertilized egg, through embryo, into adult. We can study the division of the parent cell into two, then four, then eight, and so on, until vast numbers of cells gather, then differentiate themselves into unique structures or tissues that will assume specialized functions.

Multicellular organisms comprise two large kingdoms: plants and animals. The distinction between these kingdoms is not rigid. They have common origins and possess many similar qualities. Except for the unicellular plants/animals, as described in the previous sections, plants are distinguished from animals by the absence of locomotion and by some different metabolic processes.

About 350 million years ago multicellular plants emerged from the sea and invaded dry land; they gradually developed roots to fix themselves into earth and stems to support the broad leaves that soaked up the sunlight. The early invaders were the carboniferous plants (responsible for the great coal deposits of the Carboniferous period), which reproduced without seeds. Over the ensuing years, conifers and palmlike plants with leaves resembling ferns gained prominence on the earth's surface. Approximately 120 million years ago the flowering plants appeared.

Plants are classified into several groupings that include algae, fungi, mosses, and vascular plants.

Algae

Most of the seaweeds and aquatic plants of fresh water are algae. Algae vary in size from tiny single cells, to small colonies, to massive aggregations like the giant kelp, which reaches seven hundred feet in length. Instead of the conventional root, stem, and leaf, as with ferns and seed plants, each plant in the algae group has a simple thallus, appearing as a broad, flat ribbon or shaped like leaves. Reproduction may be asexual by fission, as with primitive thallophytes; through spore formation, as with more complex thallophytes; or through male and female reproductive cells, as with the higher thallophytes.

Fungi

Multicellular fungi constitute a large family of plants, including bunts, molds, mildews, mushrooms, morels, yeasts, puffballs, truffles, rusts, and smuts. Parasitic fungi attach themselves to plant and animal hosts. Saprophytic fungi subsist on dead plants and animals or on animal excreta, thus helping to decompose organic matter and feed it back into the soil. Not so useful are the fungi which cause the dry-rot that is so destructive to wooden structures. Some fungi (*Ascomycetes*) join other organisms to live in symbiosis with them. For example, certain fungi amalgamate with particular species of algae to form the plants called *lichens*. Others combine with the roots of seed plants in what is termed a "mycorrhiza." Fungi are usually classified in three groups: (1) *Algae fungi*, which under a microscope resemble ordinary algae, are composed of a tangle of slender filaments called *hyphae;* common among these are the black molds seen on stale bread and berries. (2) *Sac fungi* have hyphae with cross-walls, creating sacs; these sacs contain spores and are sometimes covered with protective jackets known as *ascocarps*. The random distribution of ascocarps over the microscopic field yields striking photographs. (3) *Basidium fungi* are club-shaped with short stalks bearing spores at their tips; these include the rusts such as wheat rust, the smuts that attack cereals, and common toadstools and mushrooms with their many edible and inedible species.

Mosses, liverworts, and hornworts

Almost twenty-four thousand species of mosses, liverworts, and hornworts exist. They represent a higher order of plants, which, though essentially amphibian, have adapted themselves to terrestrial life. They require a moist environment for their survival and hence are found in bogs and in shady, damp places. Mosses and liverworts do not possess real roots, nor supporting structures to lift them high off the ground; rather they attach themselves to rocks, trees, and other natural objects and assume a variety of appearances—leafless, ribbon-like, and ruffled strands; leaflike appendages attached to stems; and mats of dense carpets covering the forest floor.

The vascular plants have specialized tissues that enable them to fix themselves in soil (roots), to stand erect (stems), and to absorb the sun's rays for photosynthesis (the spreading leaves). The supporting tissues throughout these plants act as conduits for nutritive materials. Thus certain vessels transport water and minerals from the roots to the leaves, while sieve cells bring nutritive substances from the leaves to other parts of the plant. Most primitive of the vascular plants is the subdivision *Psilopsida*, which has only a few species, such as the whisk fern. This group is mostly tropical and resembles plants that existed 420 million years ago. The next subdivision comprises the eleven thousand species of club mosses (*Lycopsida*), now small in size, but whose extinct ancestors three hundred million years ago grew to heights of over one hundred feet. The third subdivision is the horsetails (*Sphenopsida*) composed of only twenty-five species, most of which now exist only as fossils. The fourth subdivision (*Pteropsida*) is the largest, with over three hundred thousand species, and is universally distributed. It has highly developed root, stem, and leaf systems—easily observed under the microscope. *Pteropsida* fall into three classes. The first are the ferns, with ten thousand species, characterized by huge leaves in beautiful patterned clusters. The second class, with 450 species, consists of naked seed plants, such as seed ferns, conifers, cycads, and ginkos. These plants originated 250 million years ago. The third and largest class (almost three hundred thousand species) are the flowering plants, which include most of the fruits, vegetables, grasses, nuts, grains, flowers, and large trees. The distinguishing feature of these plants is that they reproduce by the process of cross-pollination. An infinite variety of subjects awaits the photomicrographer in the versatile designs of the roots, stems, leaves, and flowers in this category of plant life.

Wood tissues

Wood, the tough, fibrous substance in the trunks and branches of trees, is one of the most versatile materials known to man. Its combustibility makes it ideal as a fuel. For construction purposes its strength, stiffness, and hardness permit it to accommodate the severe stresses of tension, compression, and shear. Its toughness and durability are outstanding. Moreover, wood yields chemical products of great commercial importance, such as dyes, rubber, maple sugar, quinine, camphor, rosin, turpentine, flavorings, medicines. Wood is an excellent source of cellulose—important in the making of paper, rayon, and other synthetics. Treated with hydrochloric acid, wood produces crude sugars, which can be used as cattle food or which can be fermented to make alcohol. Through hydrogenation wood is converted into liquid fuels.

Viewed through the microscope, woods are among the most beautiful multicellular plant materials, revealing unusual patterns and shapes found in no others. We observe in the architecture of wood a fascinating maze of intricate vascular tissues (*xylem*) serving as the carriers of water and essential salts—or what we commonly know as sap. This life-giving substance travels from the plant's roots to the smallest branches and leaves.

The tough fibrous elements in wood form structural supports for the plant. As a tree ages, the central part dies and its ducts become plugged; the cellular protoplasm is replaced by either gums or resins. Or, in some plants, the ducts merely remain empty.

Wood is divided into two groups: softwoods and hardwoods. Softwoods (pines, spruces, larches, and so on), when sliced sufficiently thin to be illuminated by bright-field methods, reveal under the microscope slender, hollow, spindle-shaped fibers (*tracheids*) that run parallel to the long axis of the trunk. The walls of each fiber contain pits that enable the sap to course through adjacent fibers. Certain softwoods, such as the pines, also have ducts filled with resin; these ducts are shaped like tall, factory chimneys enveloped by brick-shaped cells. Stringlike rays traverse the wood, at right angles to the fibers, from the outside to the pith. These structures are composed of elongated cells that contain albuminous material, starch, sugar, or oil—luscious morsels for parasitic animals and fungi.

Hardwoods (ash, beech, oak, teak) contain relatively long, wide, water-conducting tubes (*wood vessels*) joined end to end; these tubes, when prepared as a cross-section specimen, look like pores. When hardwood is sliced longitudinally, the dense wood rays look much like ornamental grain designs. These physical characteristics may be seen in the photomicrographs of wood in this section. The thin slices of wood used as specimens were treated with a red stain and illuminated between crossed polarizers. The scattering of light through the fibers, ducts, and rays produces many fascinating configurations.

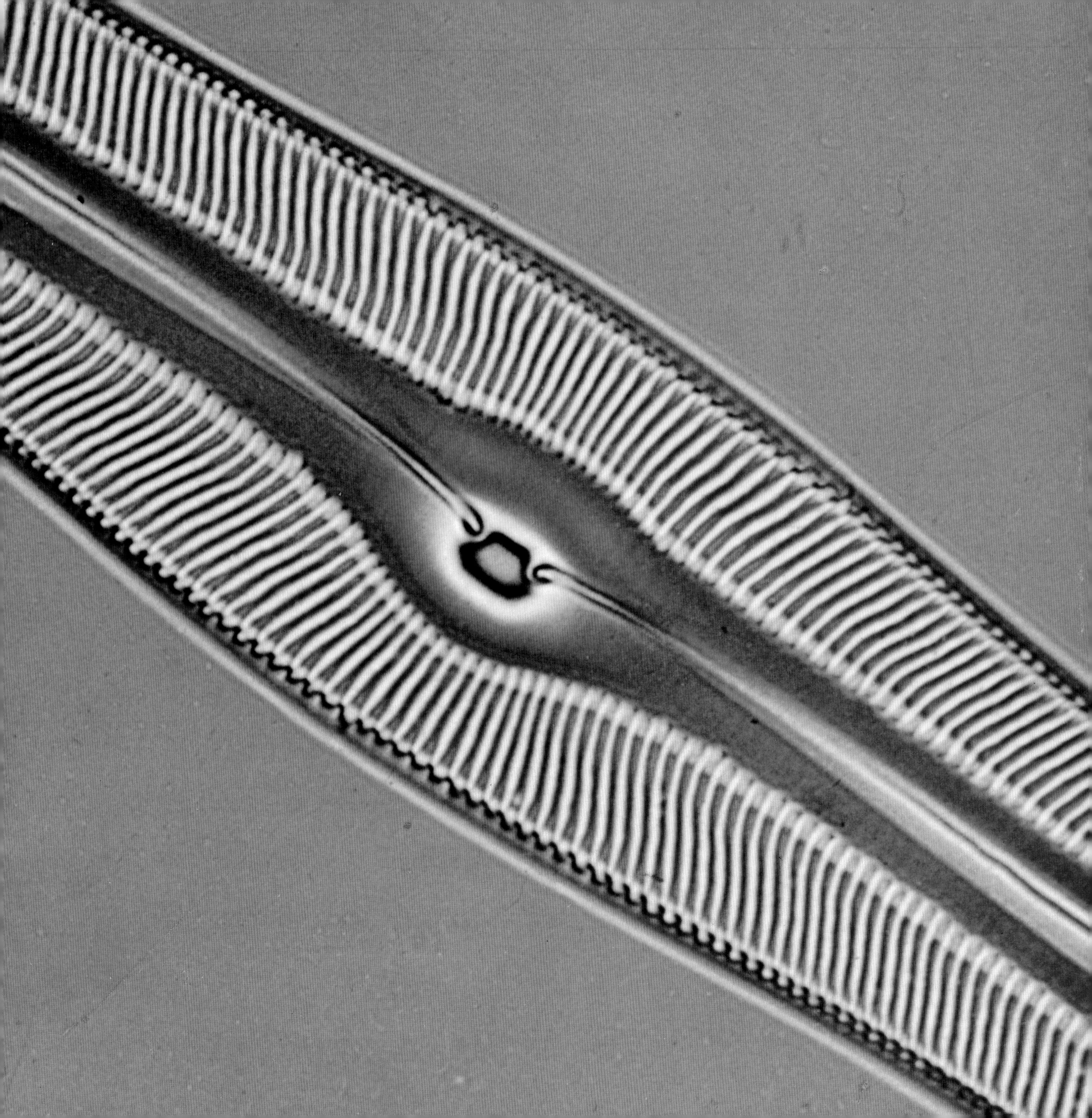

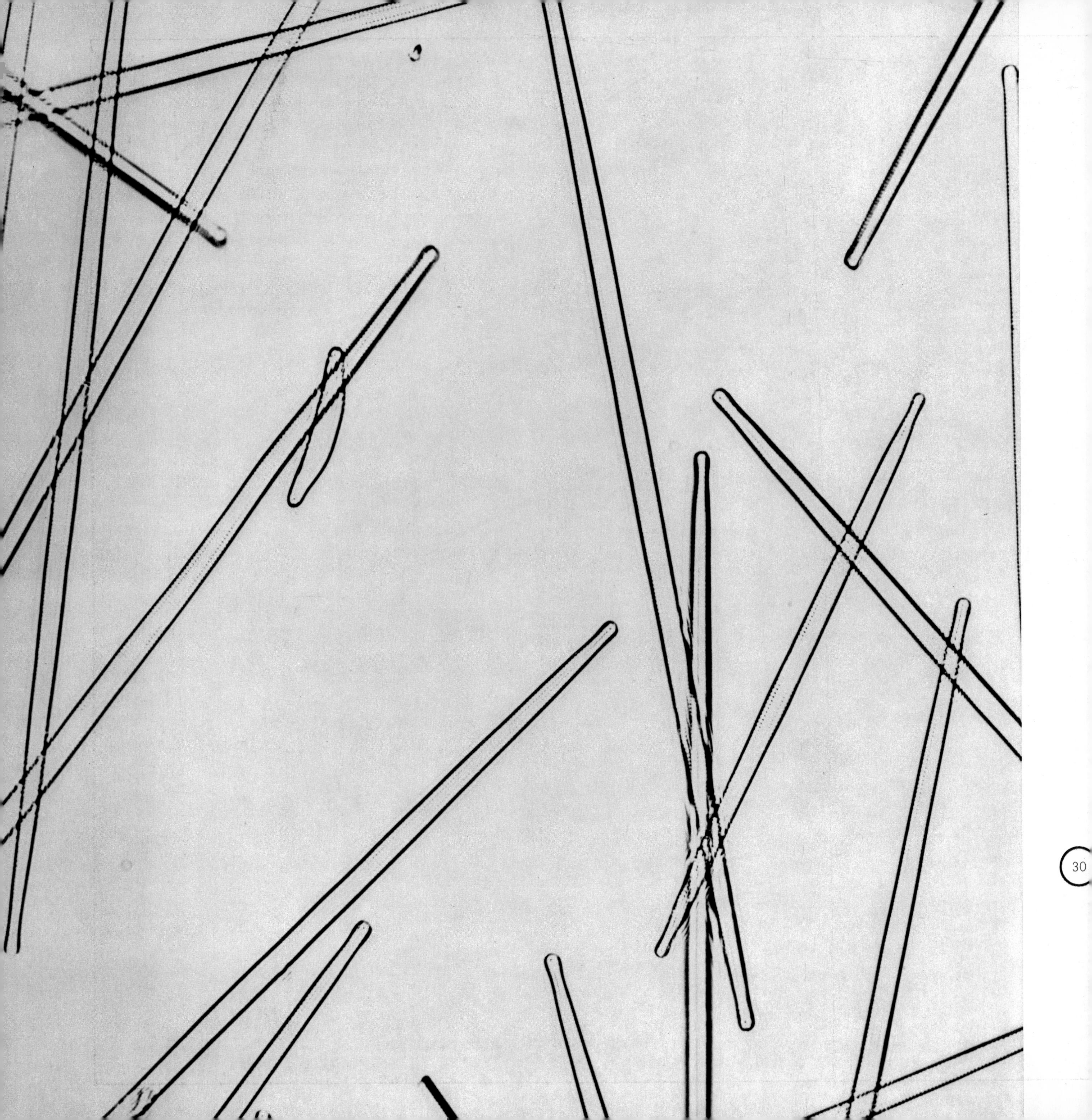
30

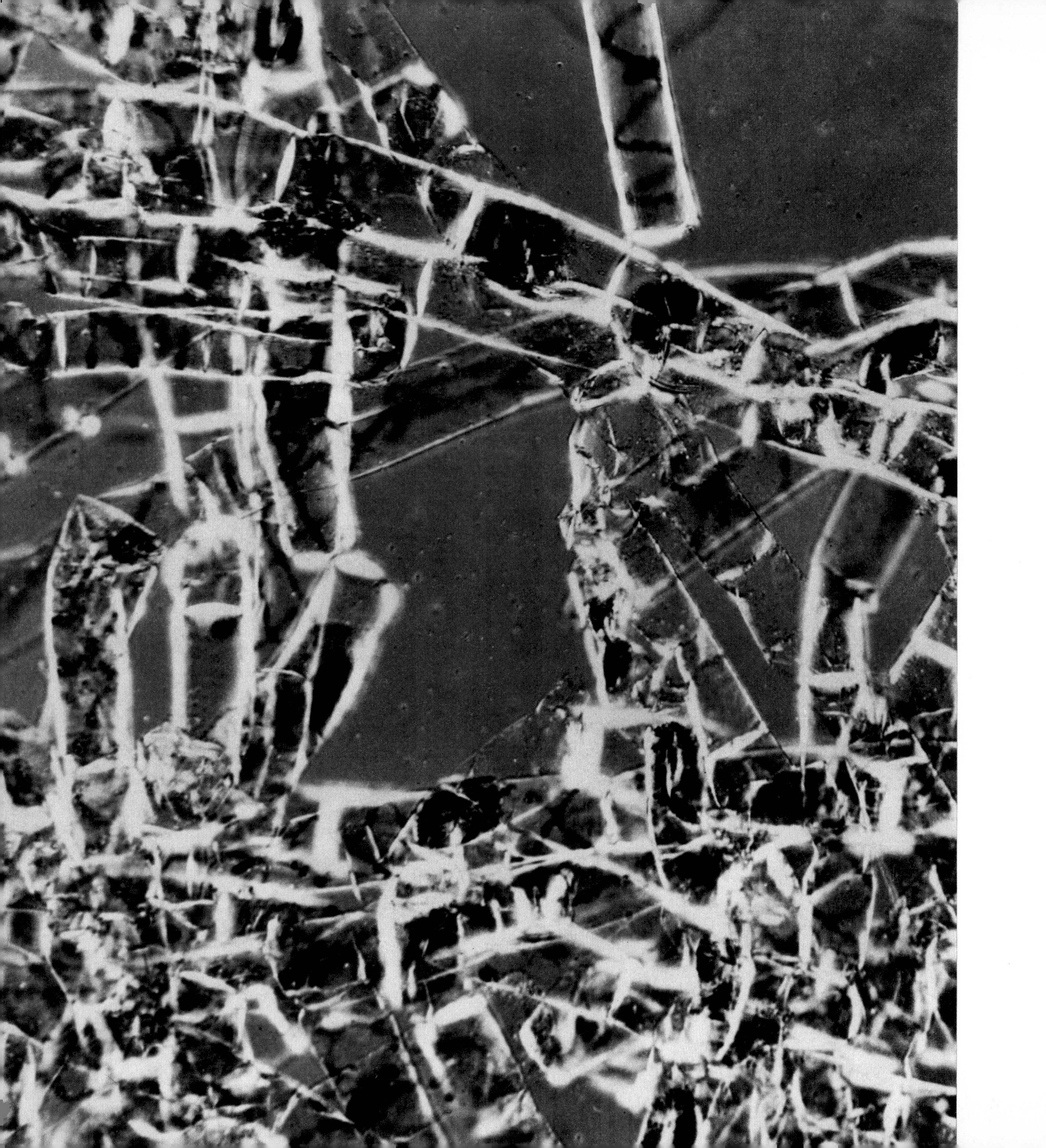

32

33

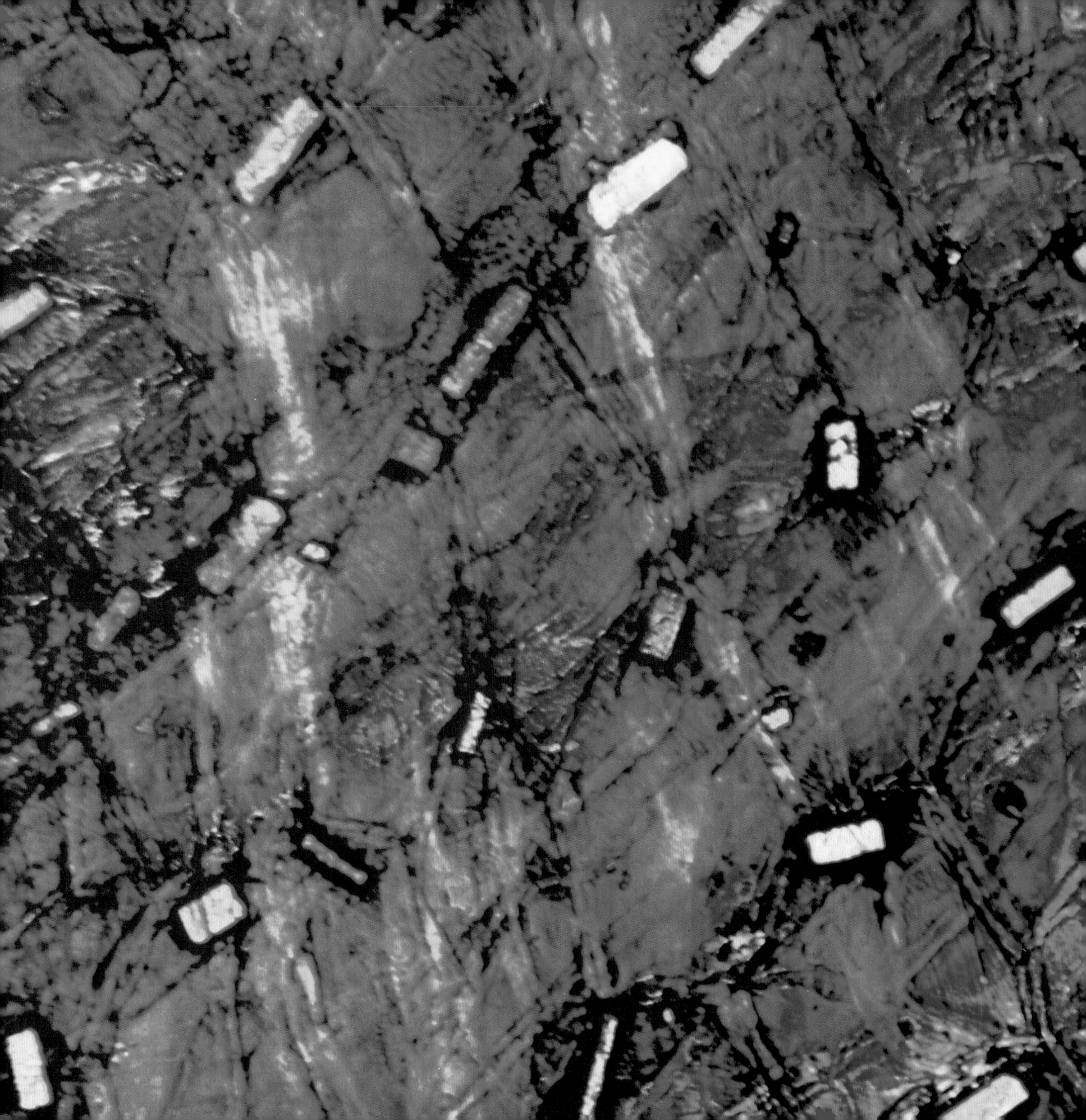

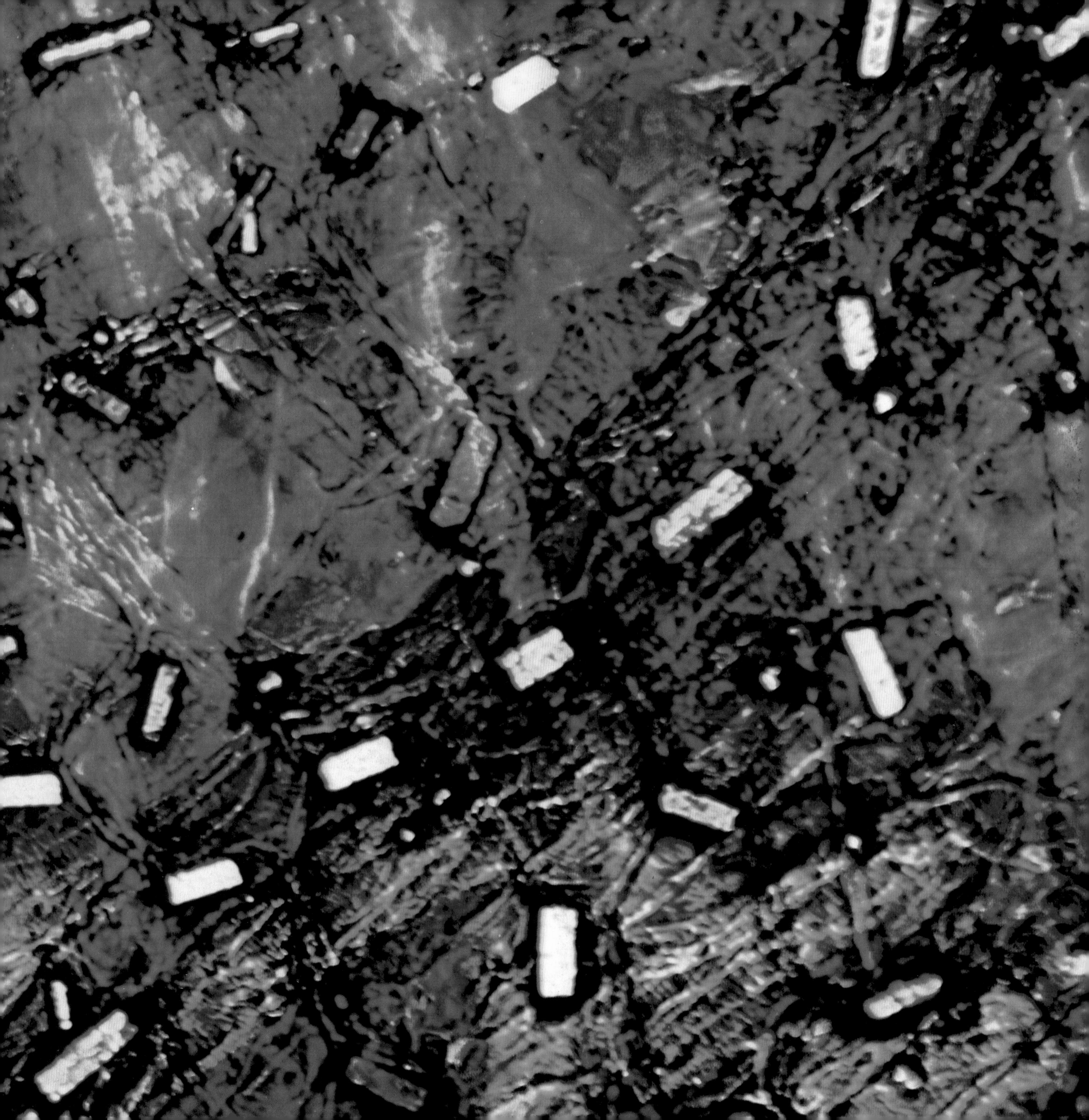

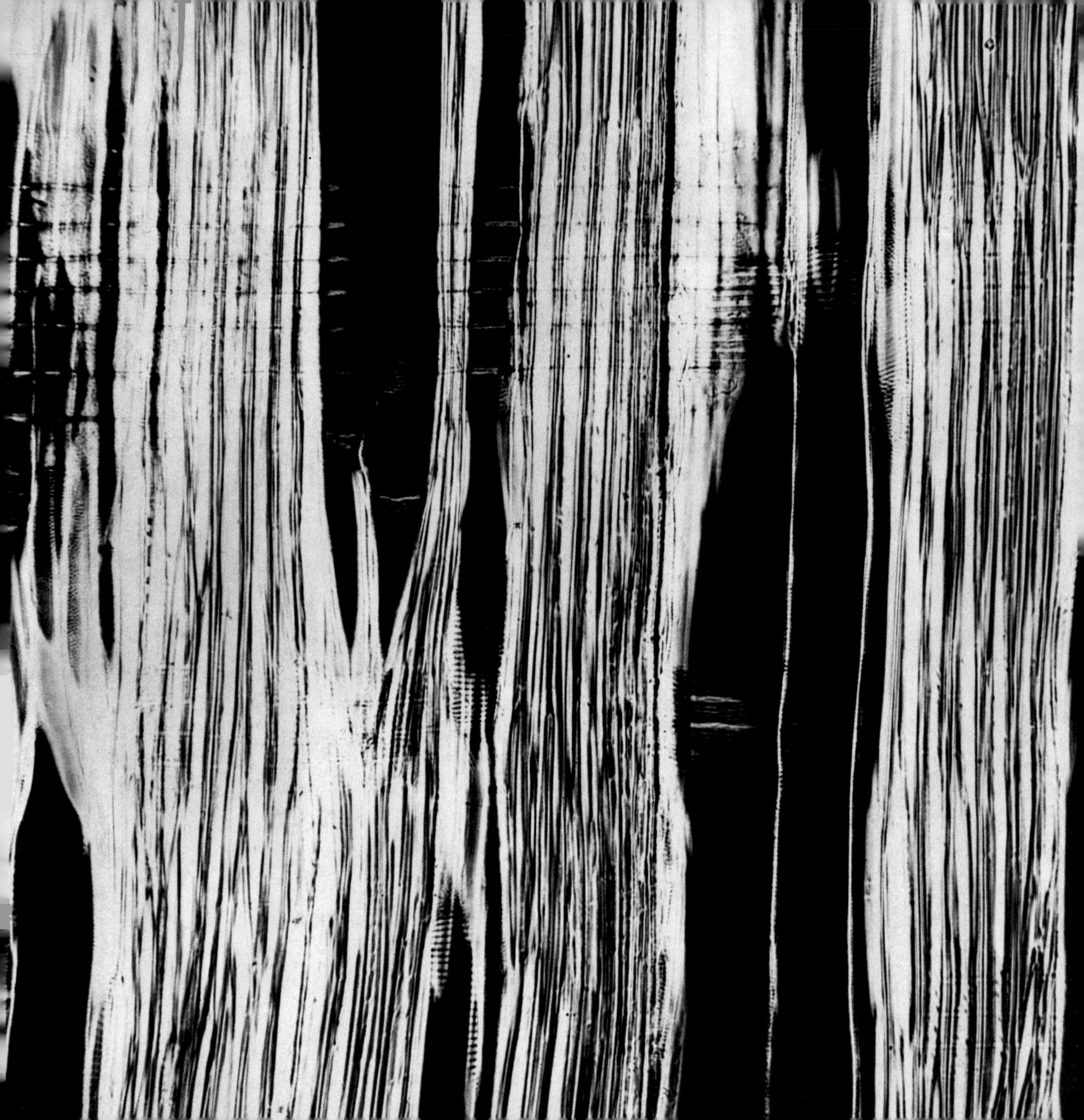

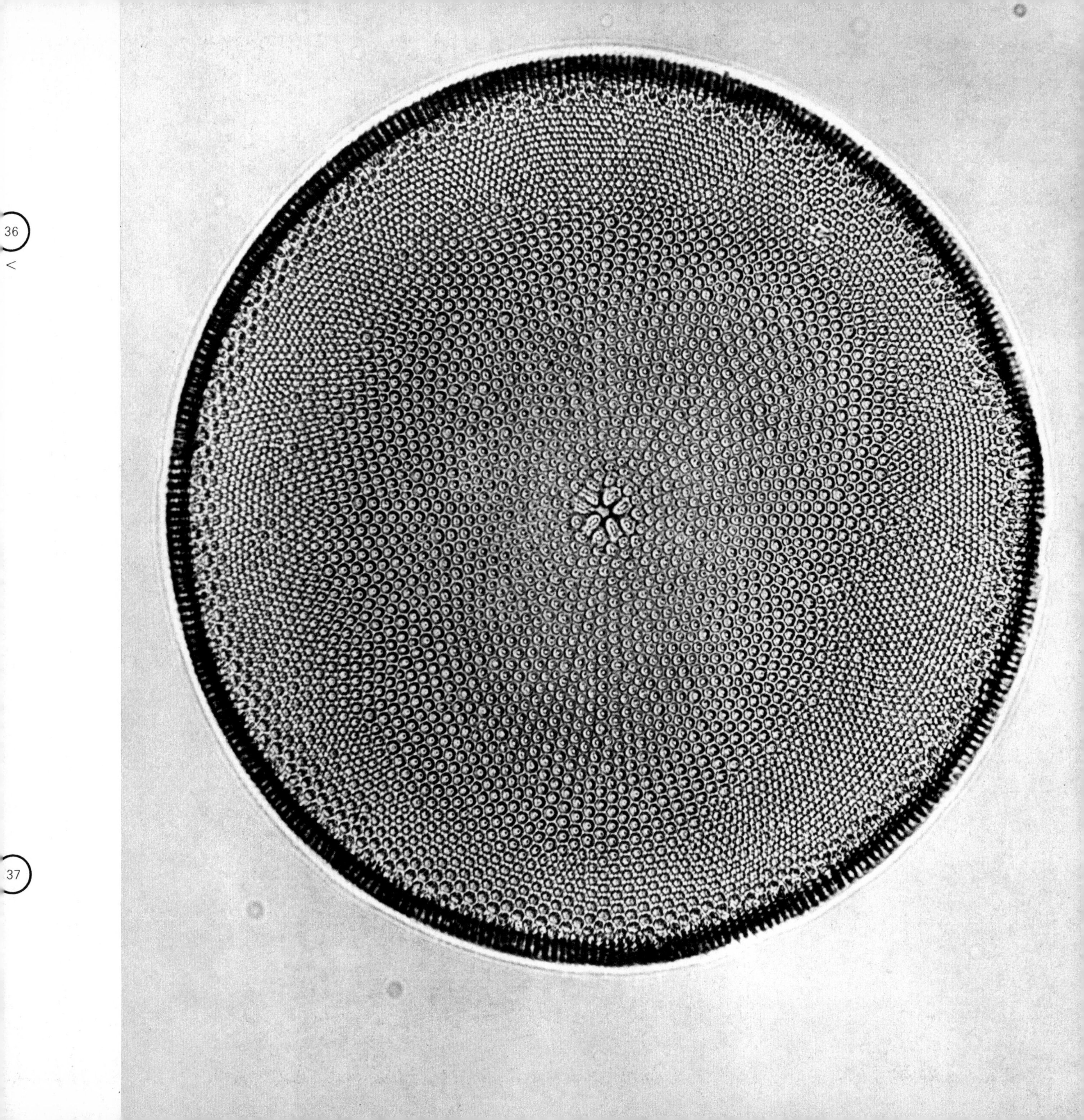
36
37

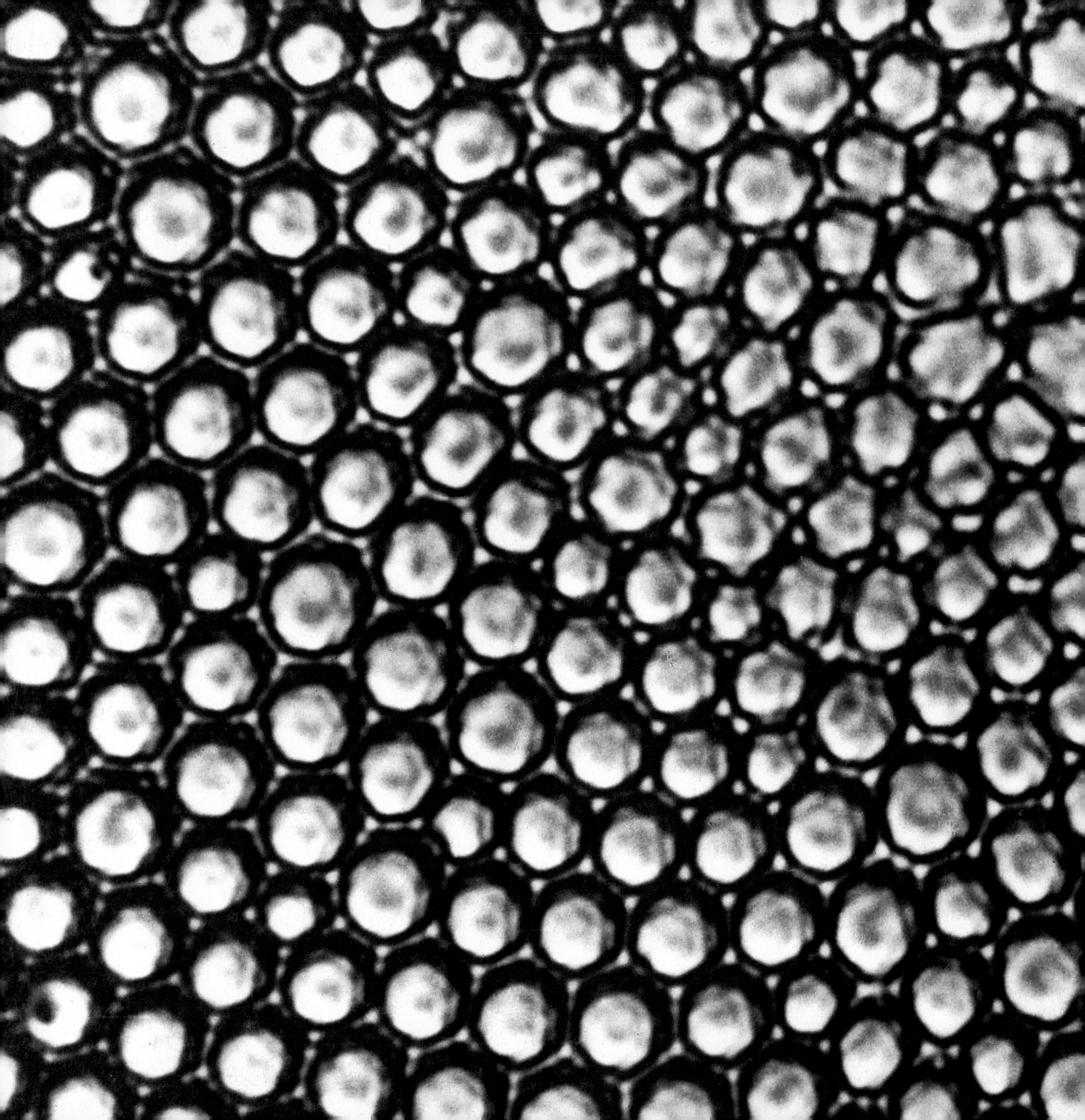

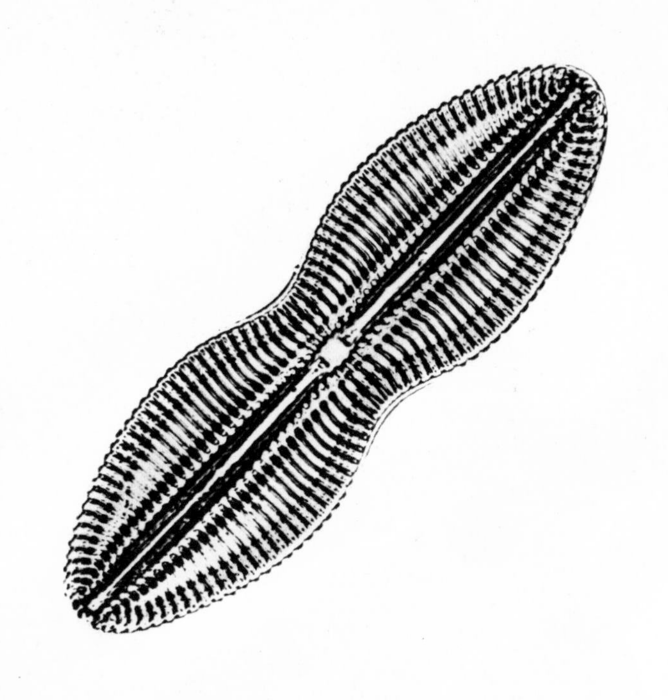

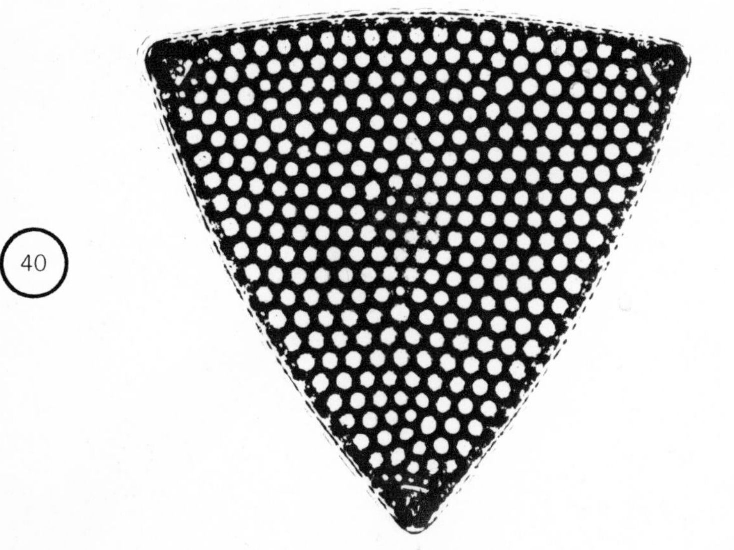

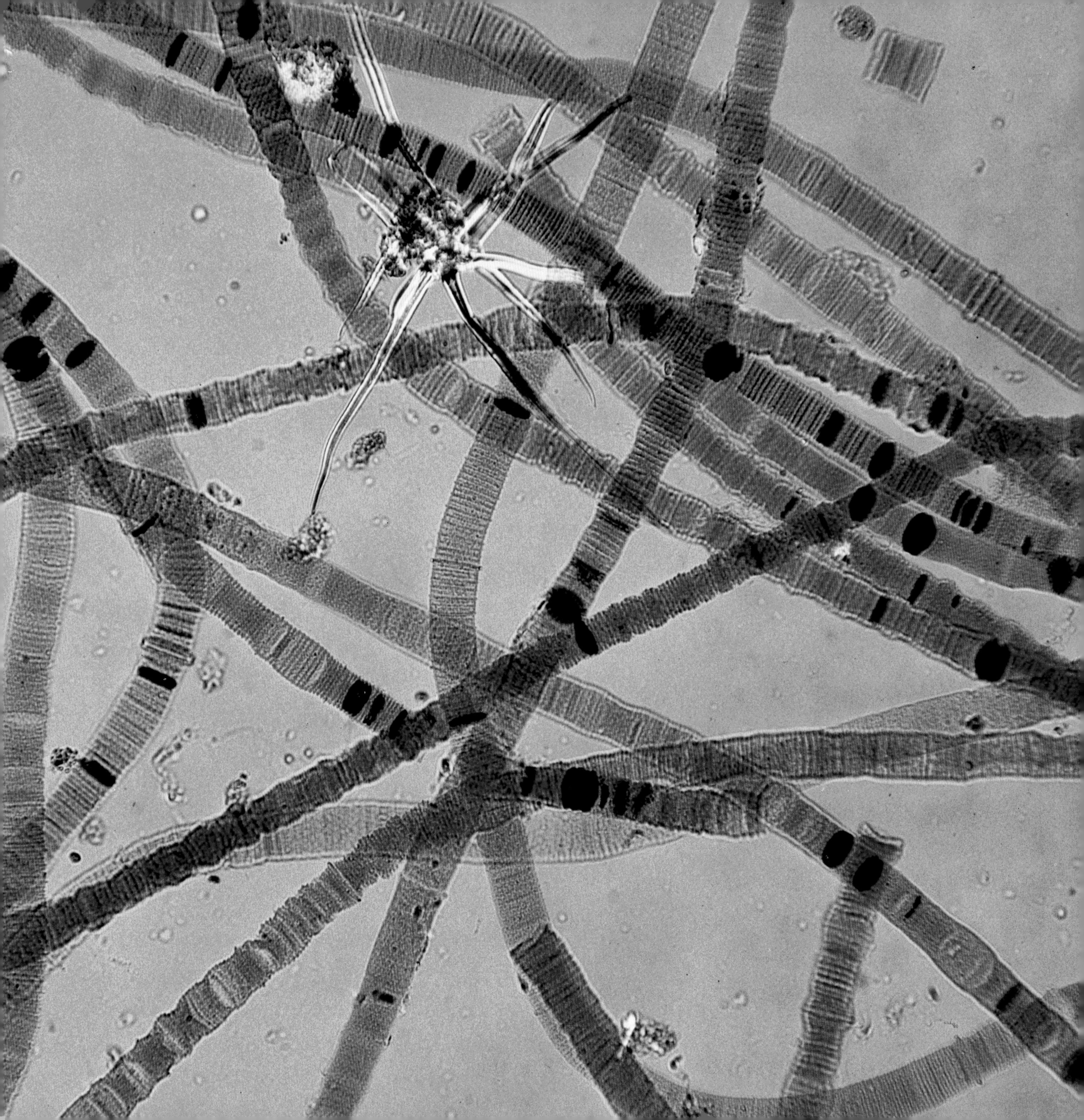

42

43

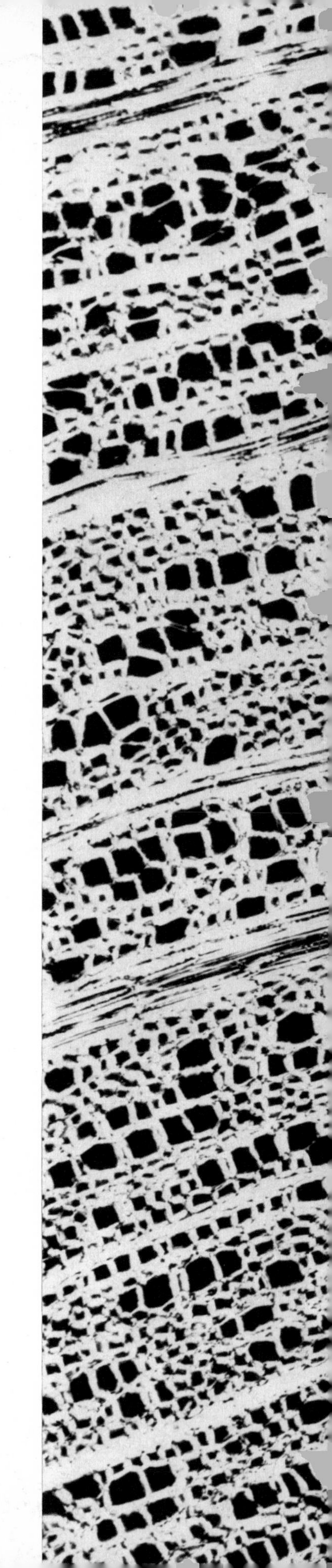

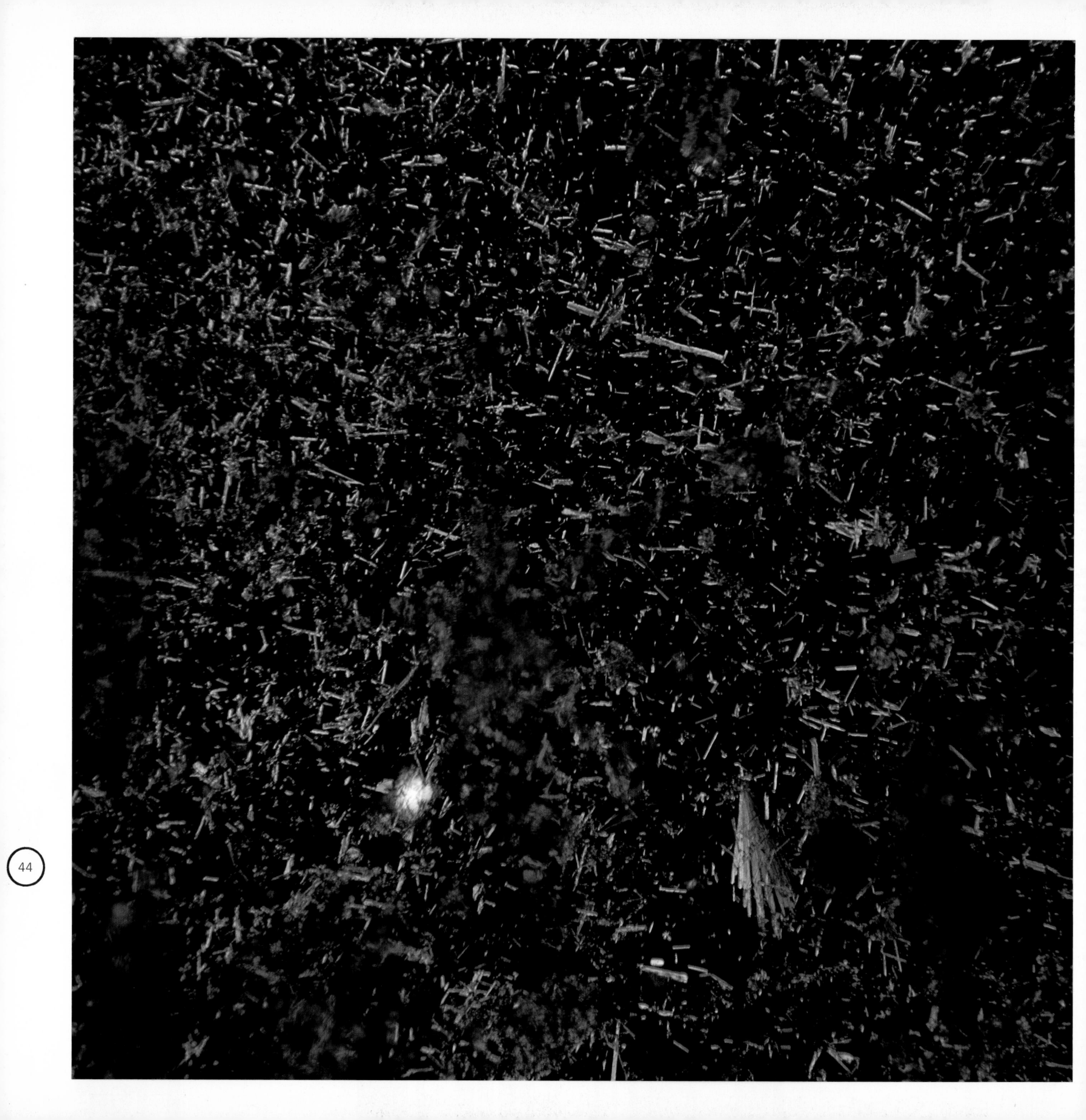

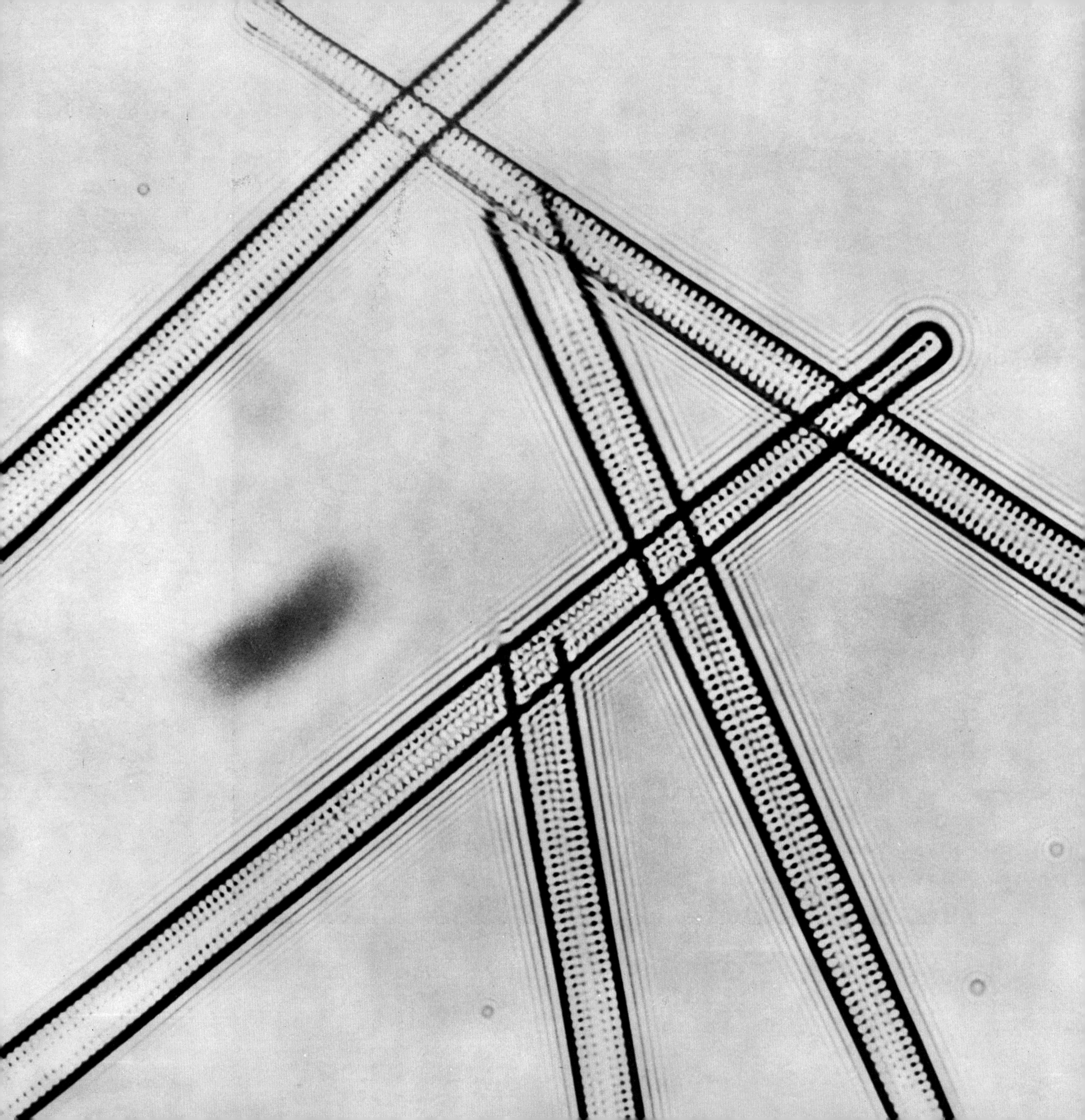

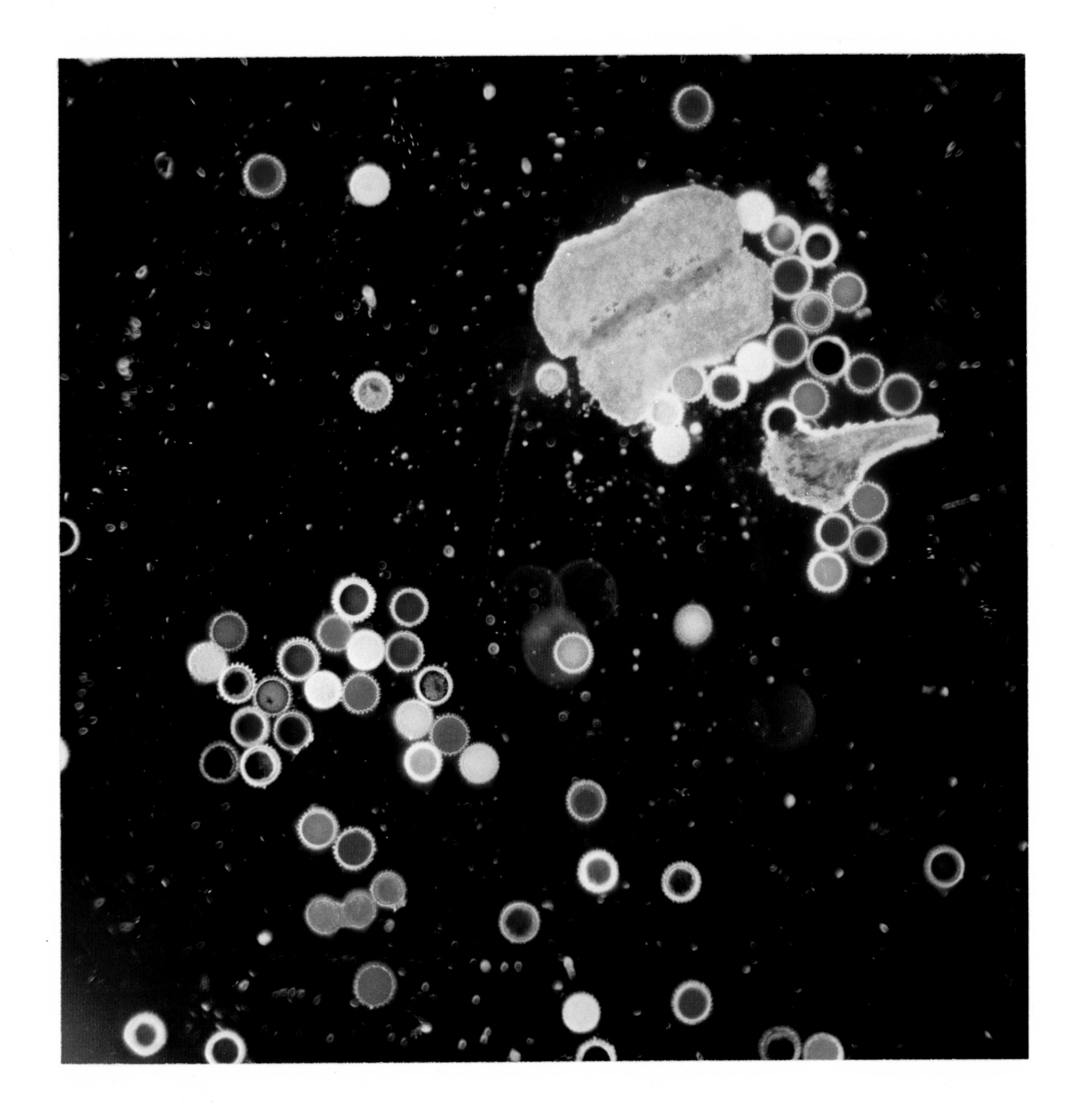

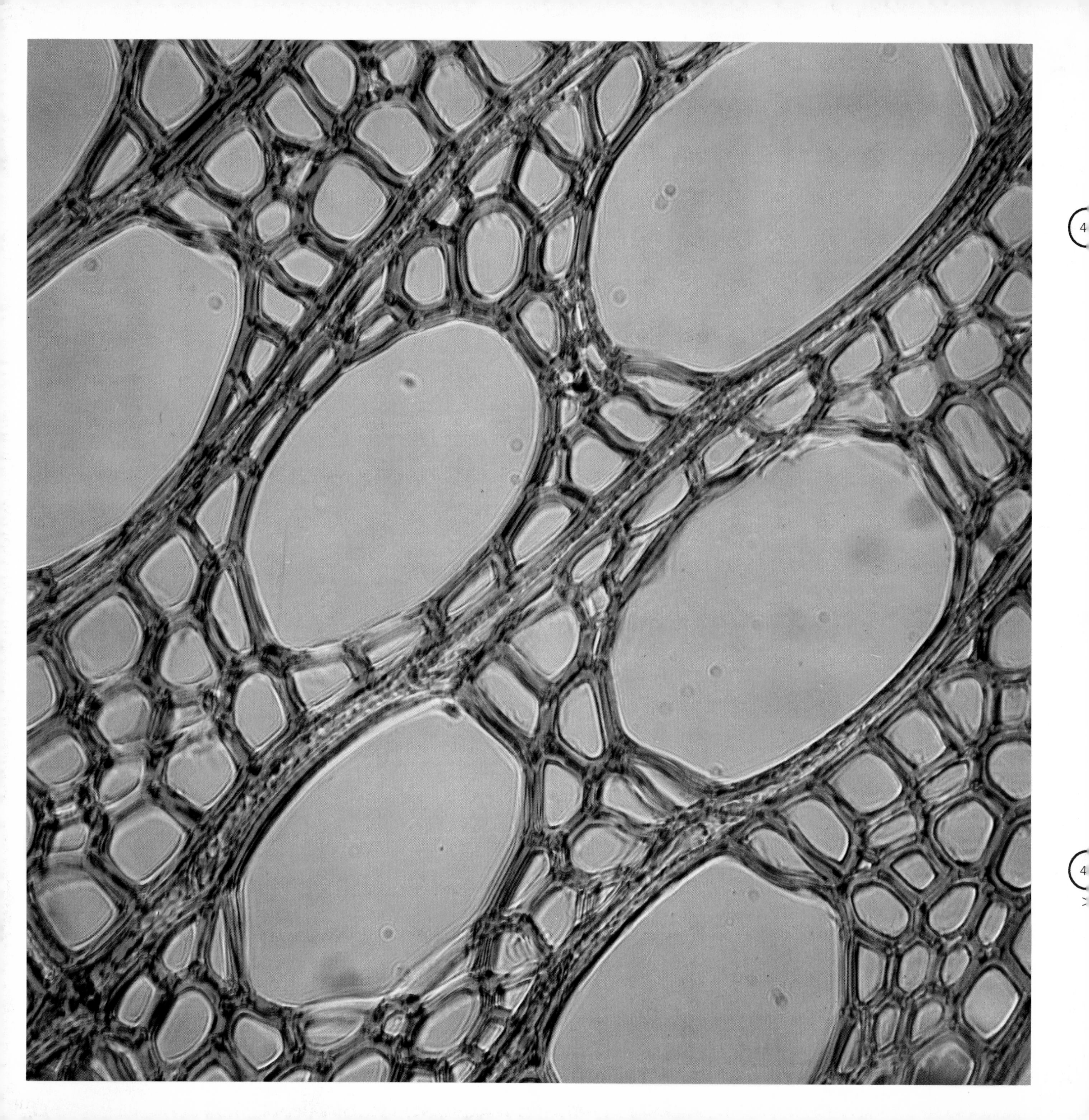

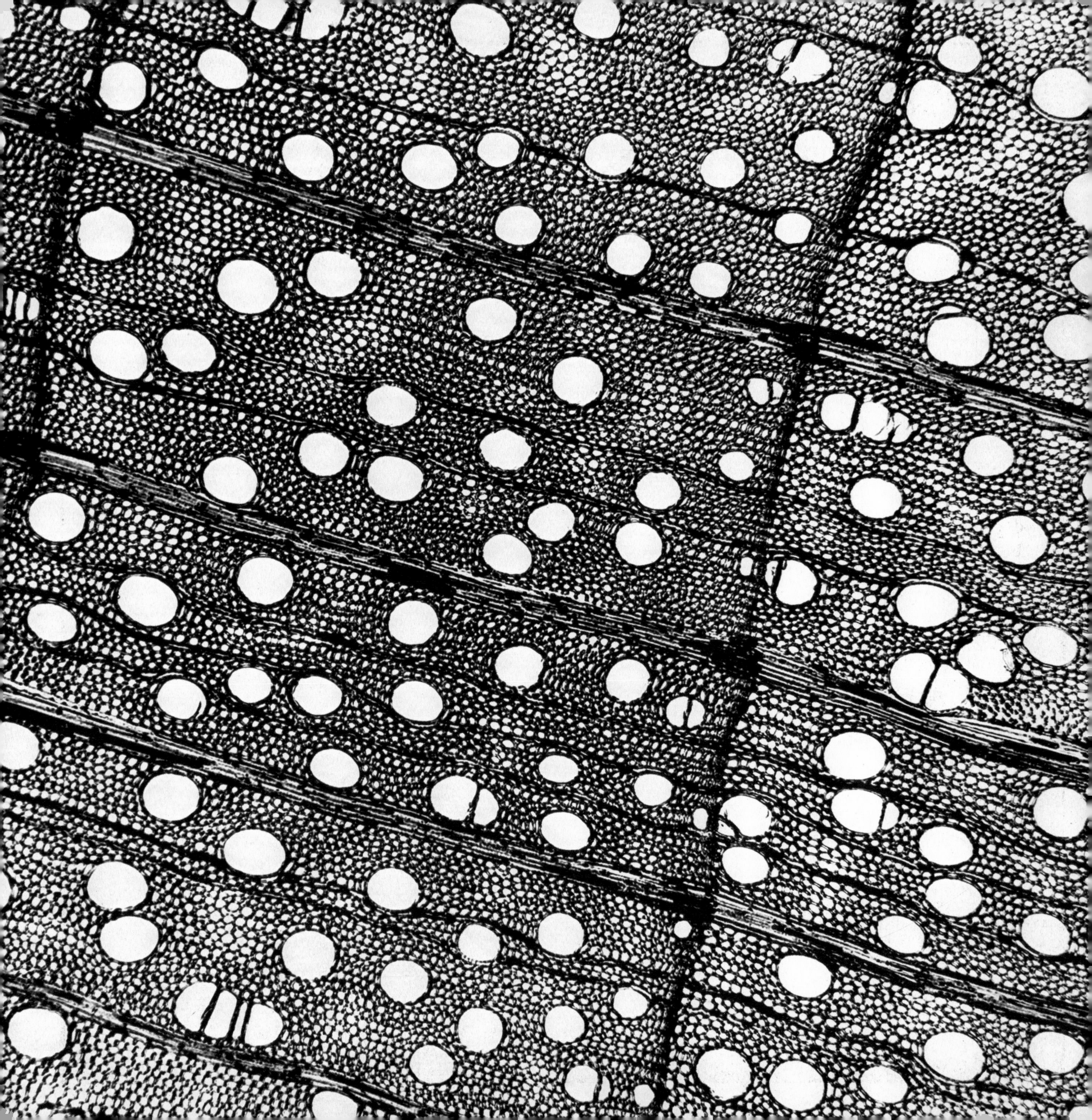

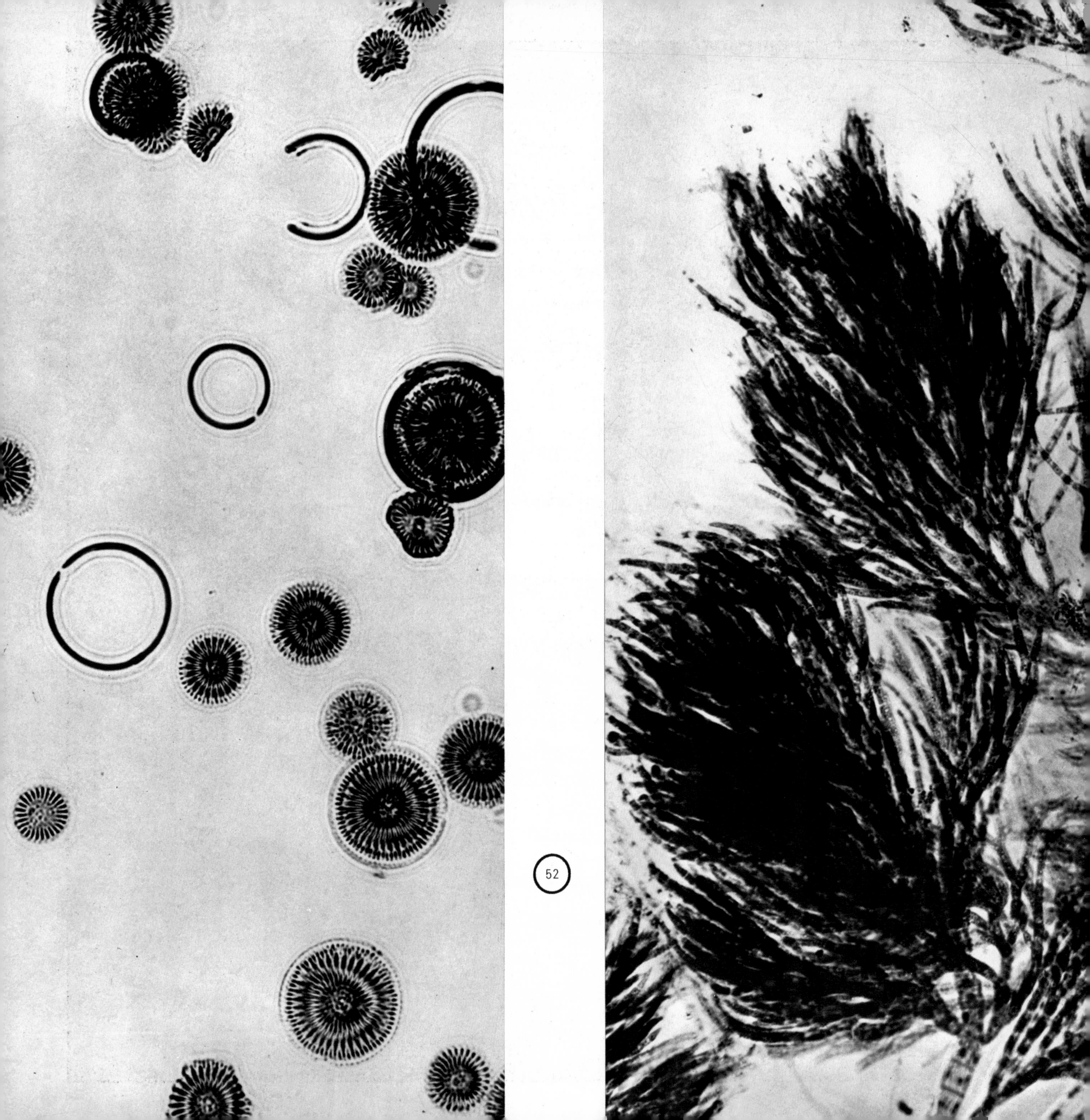

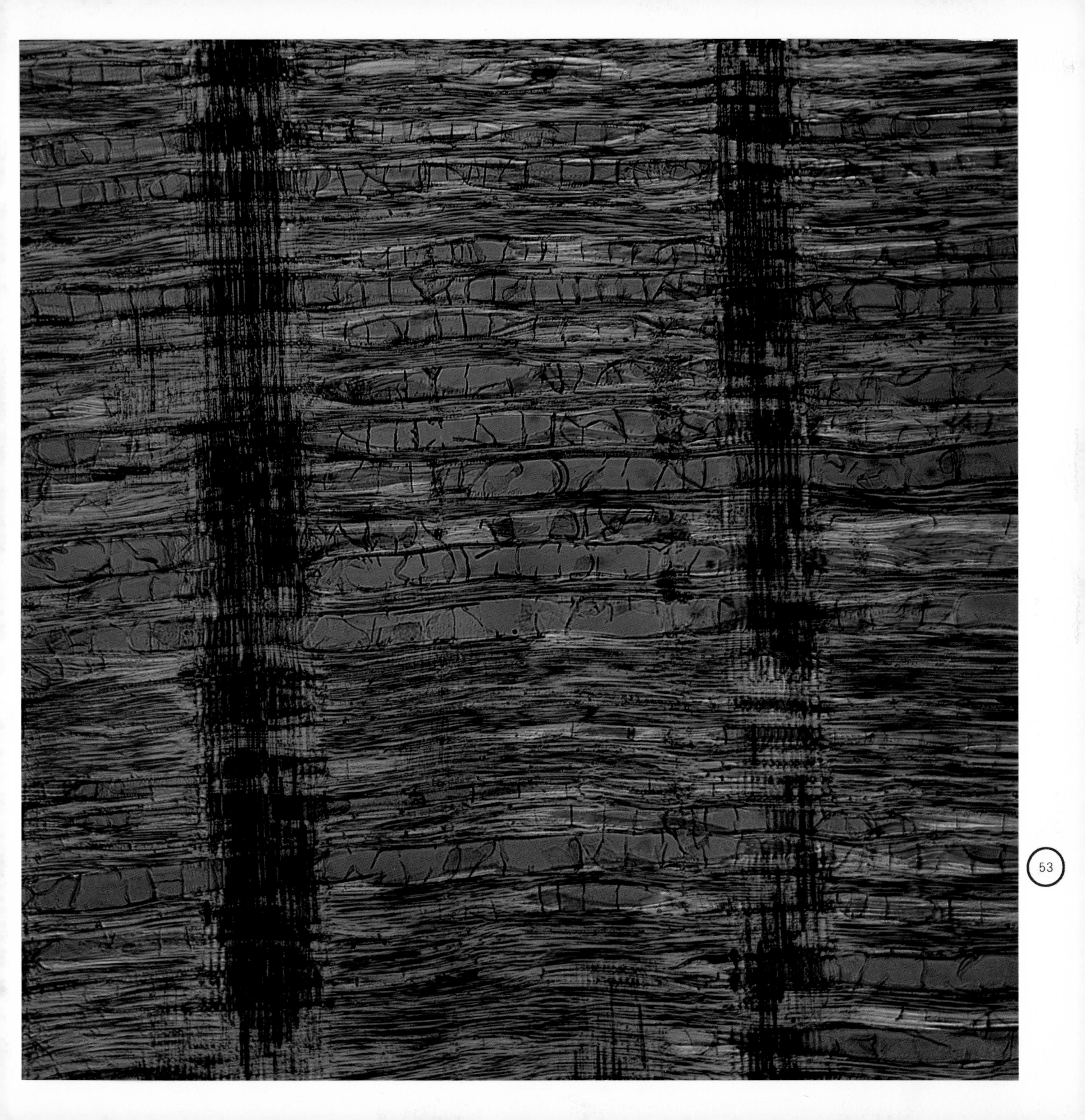

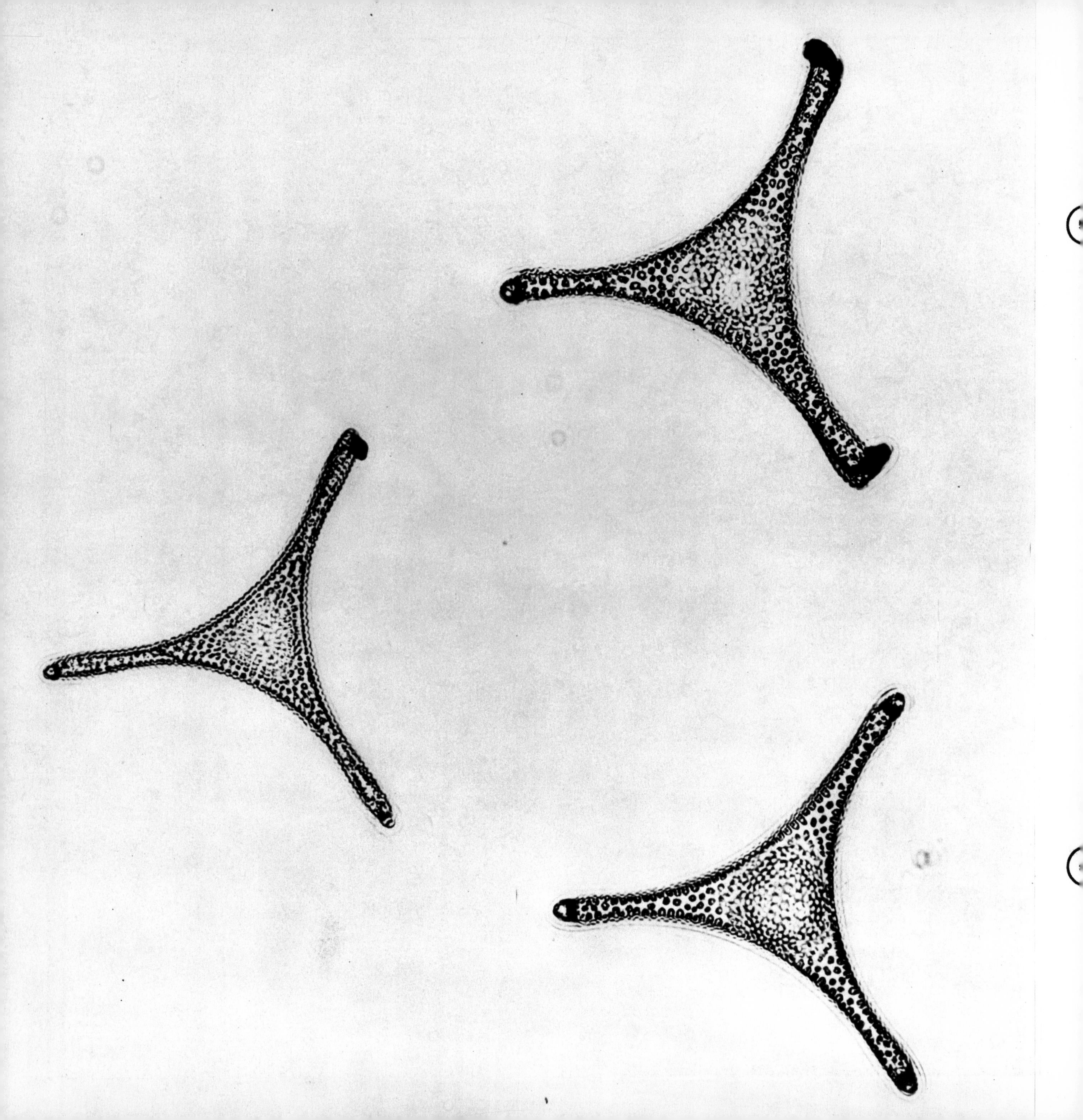

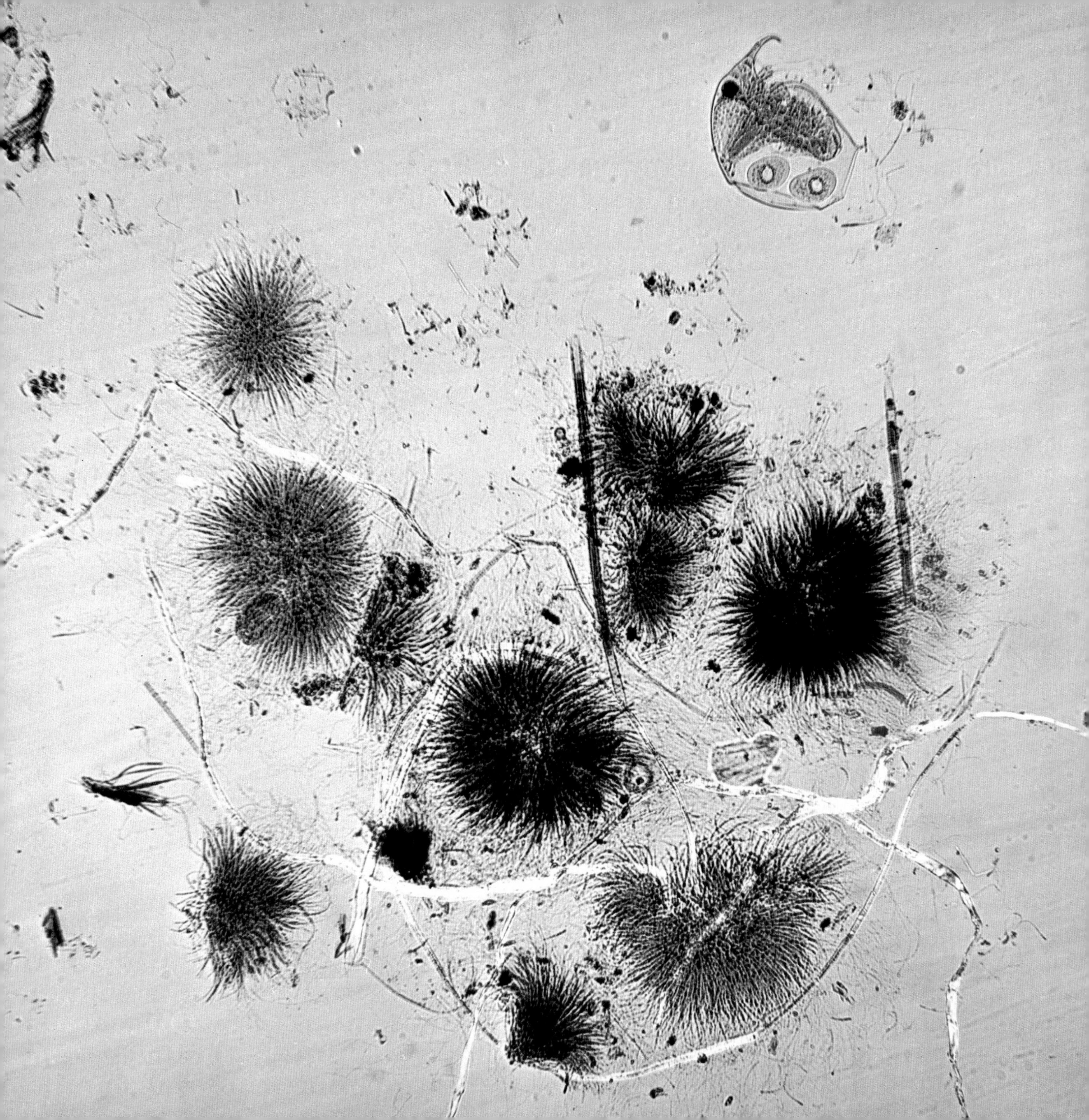

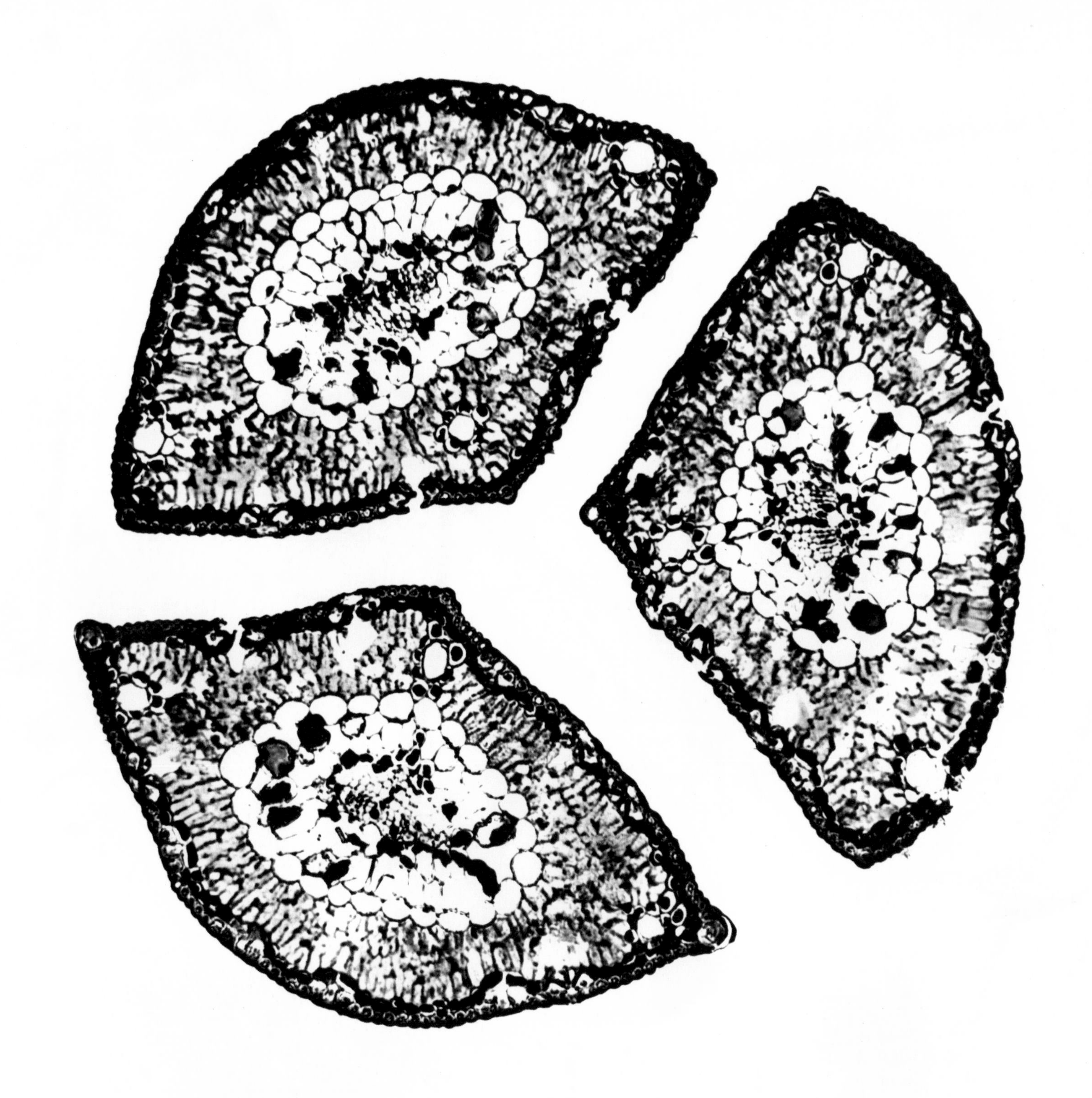

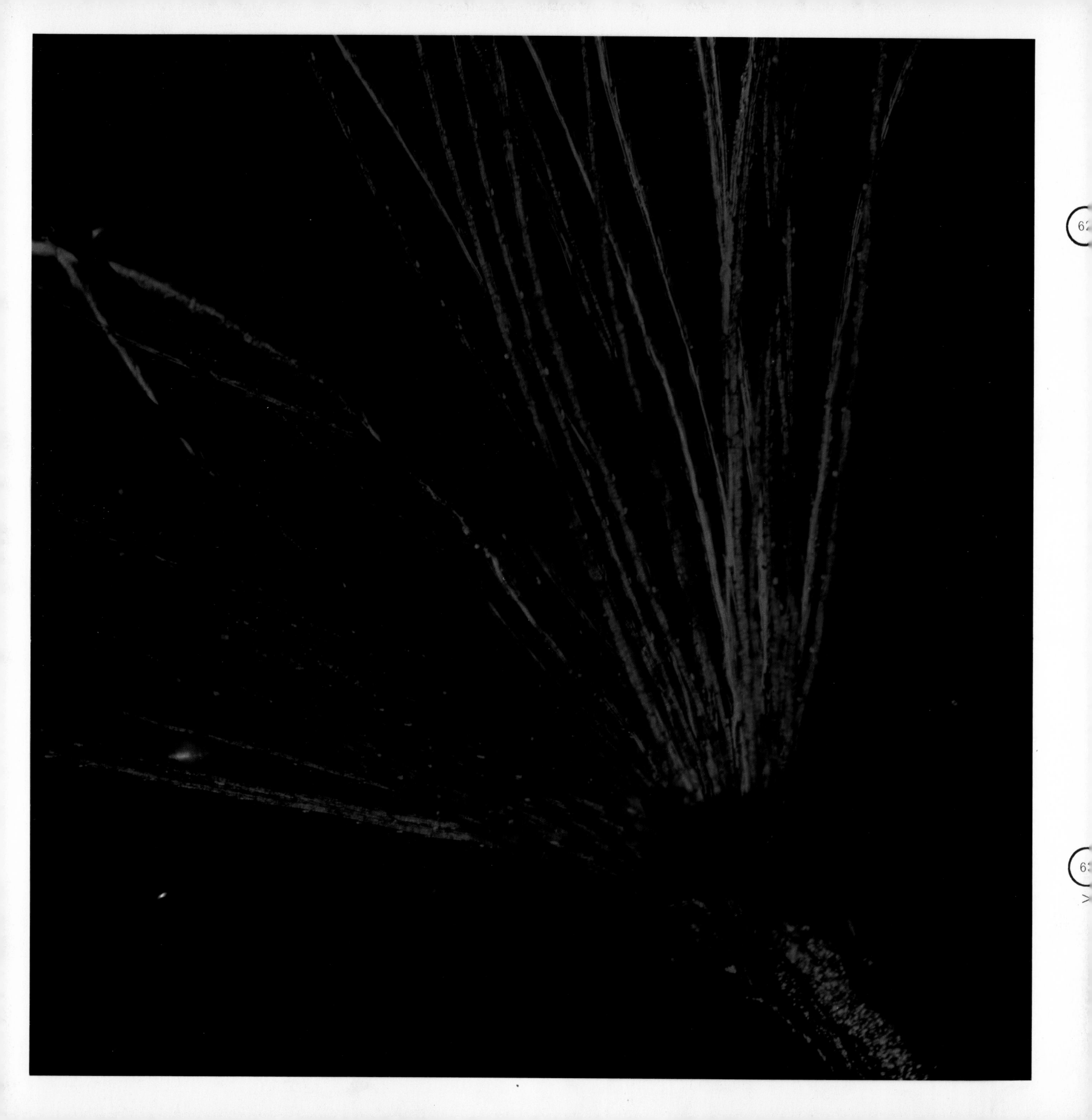

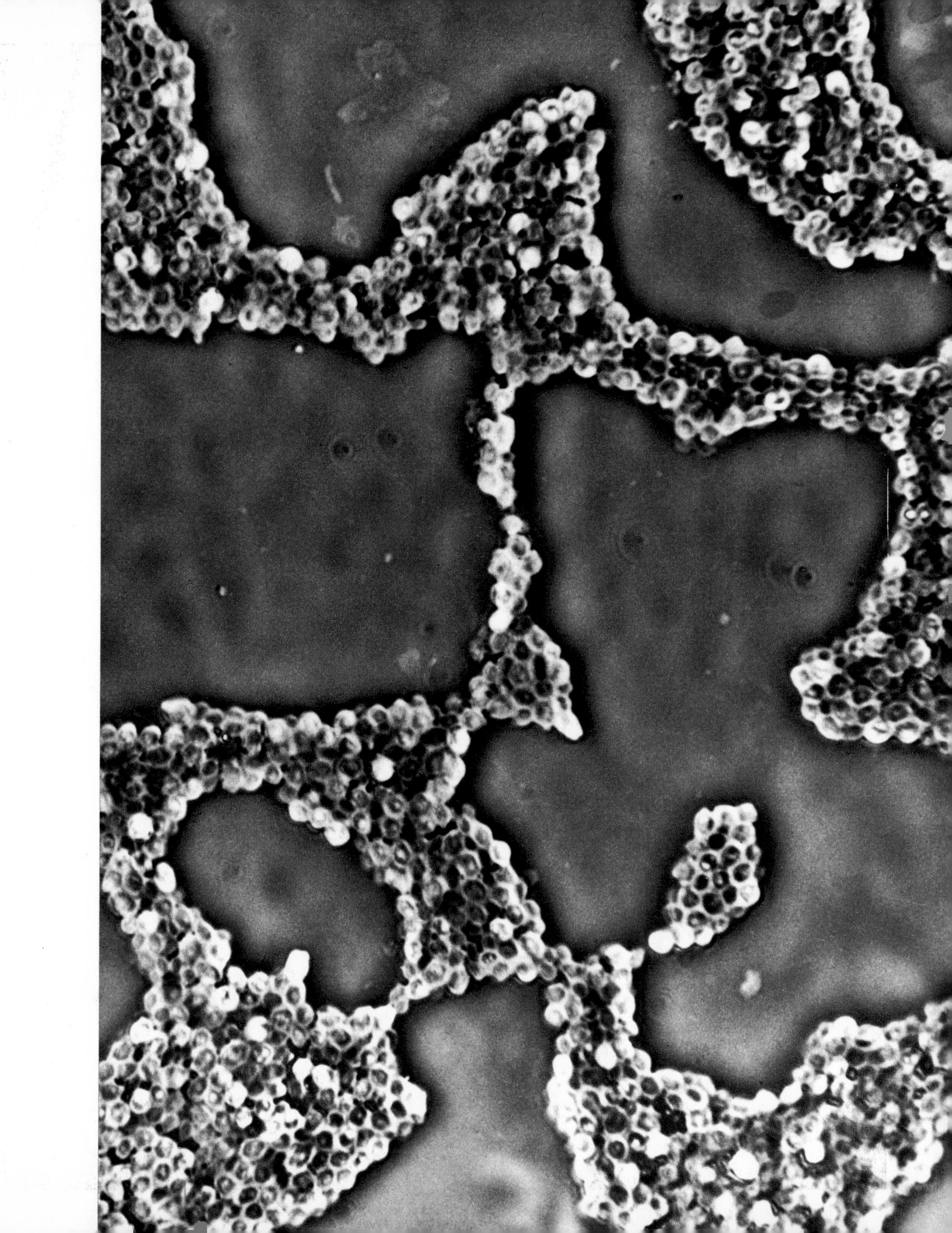

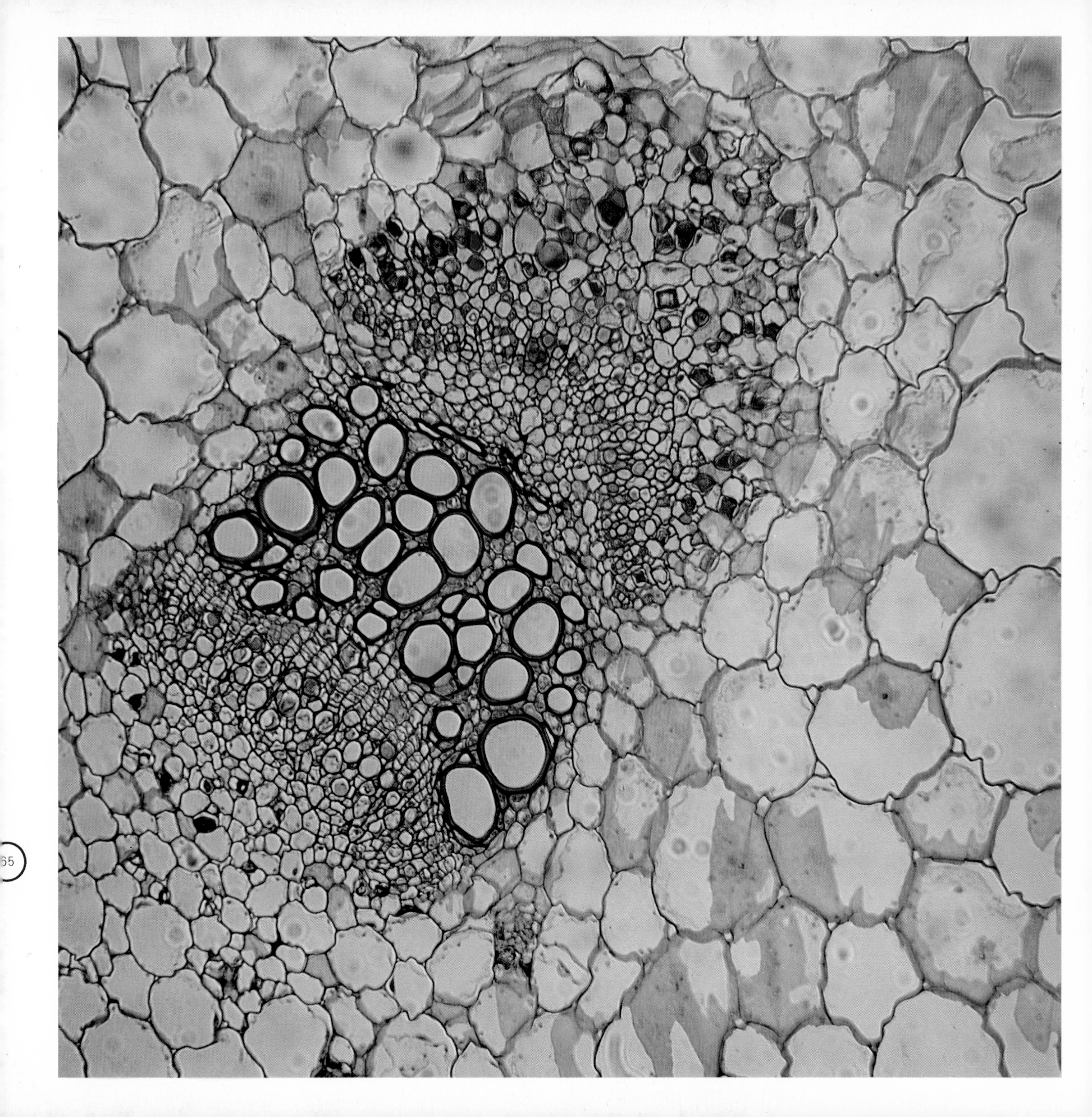

65

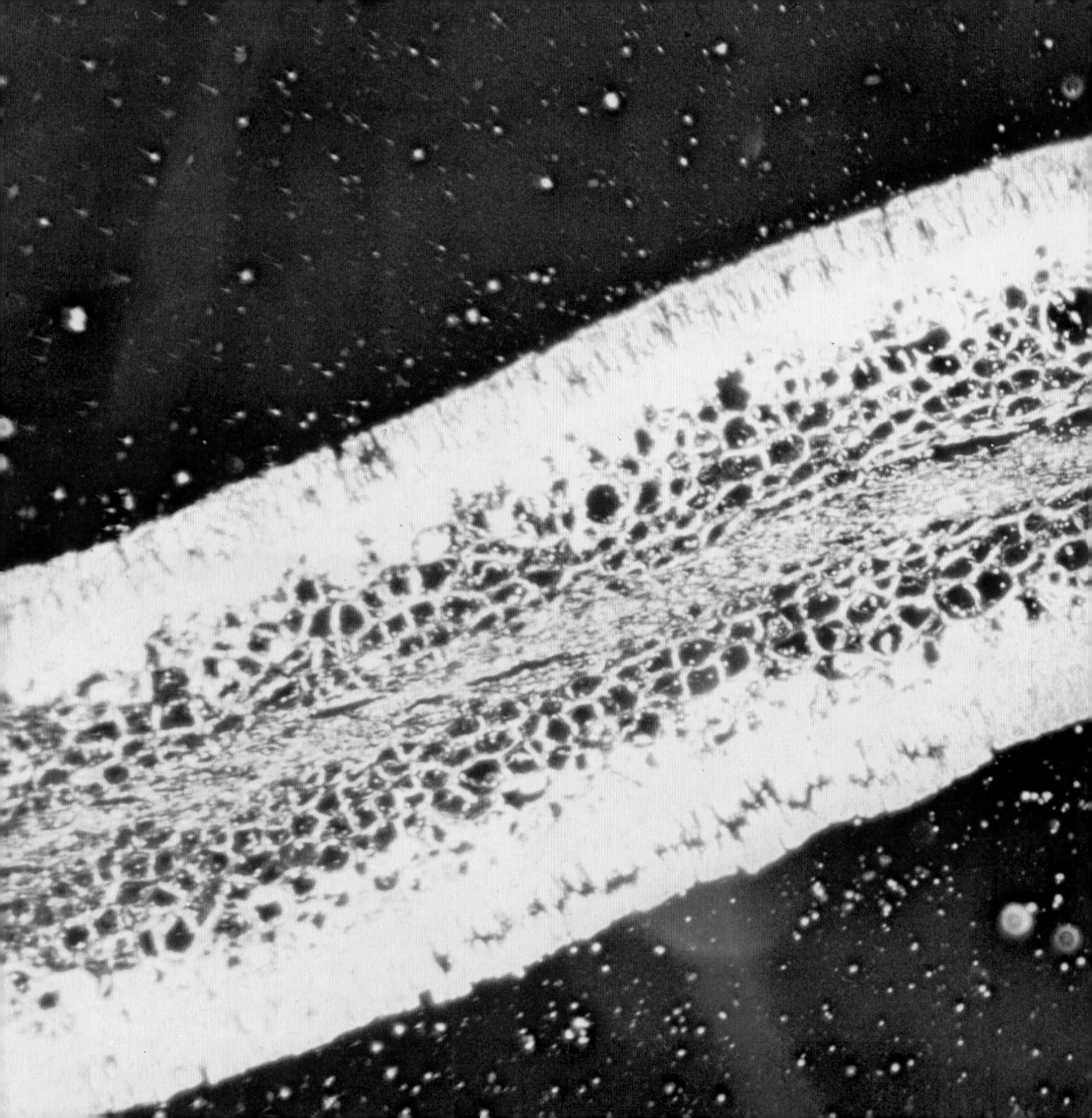

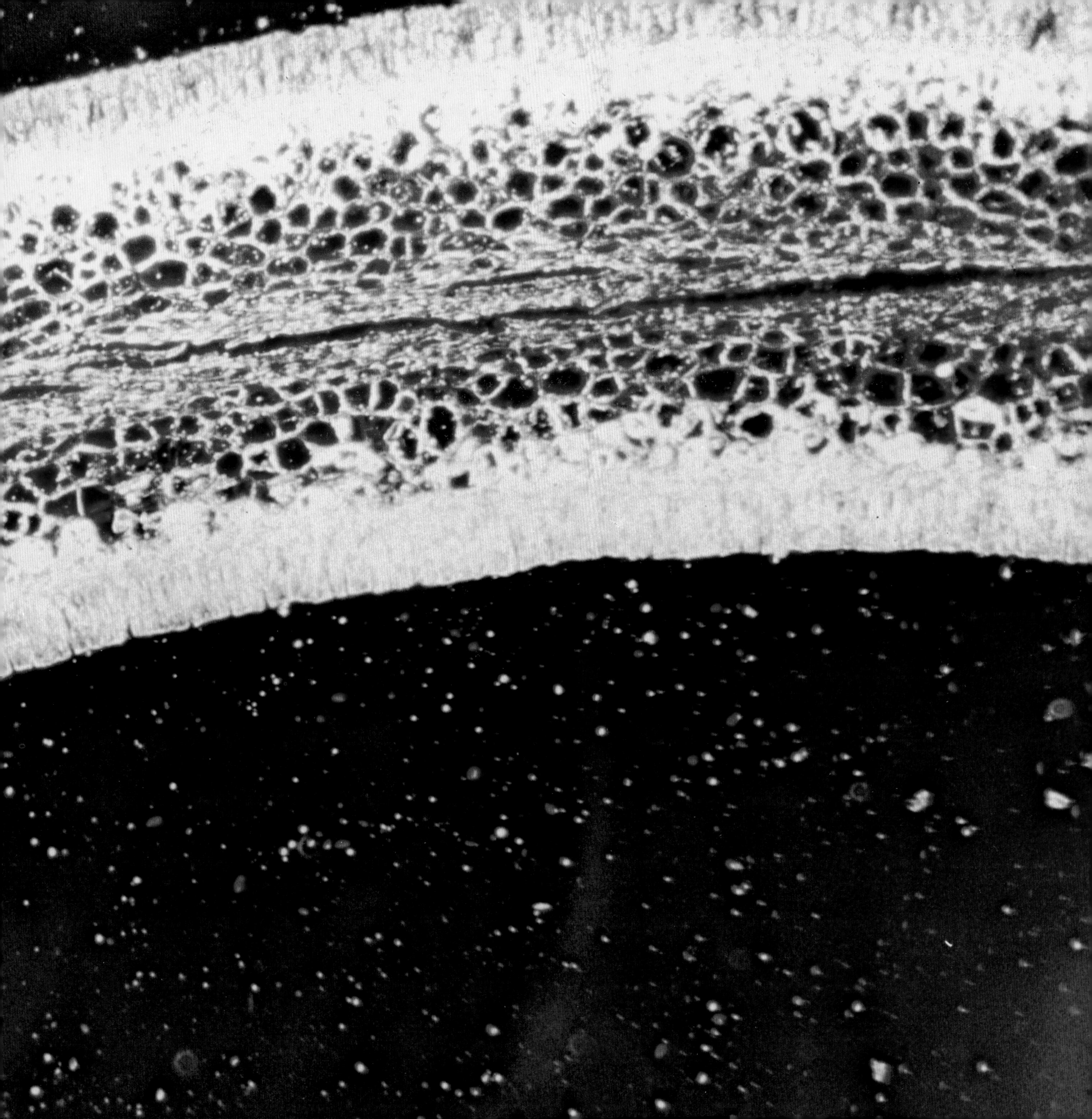

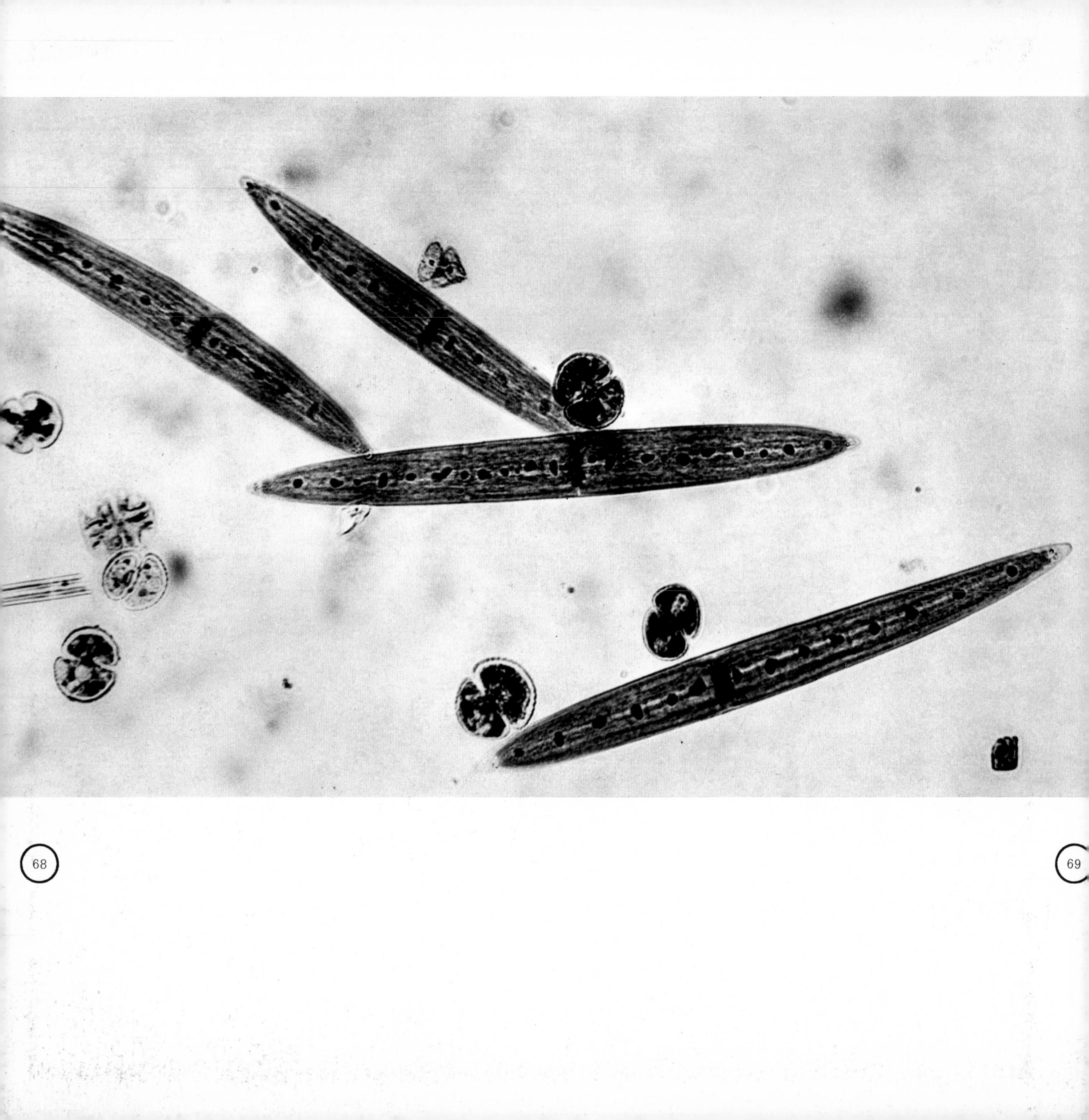

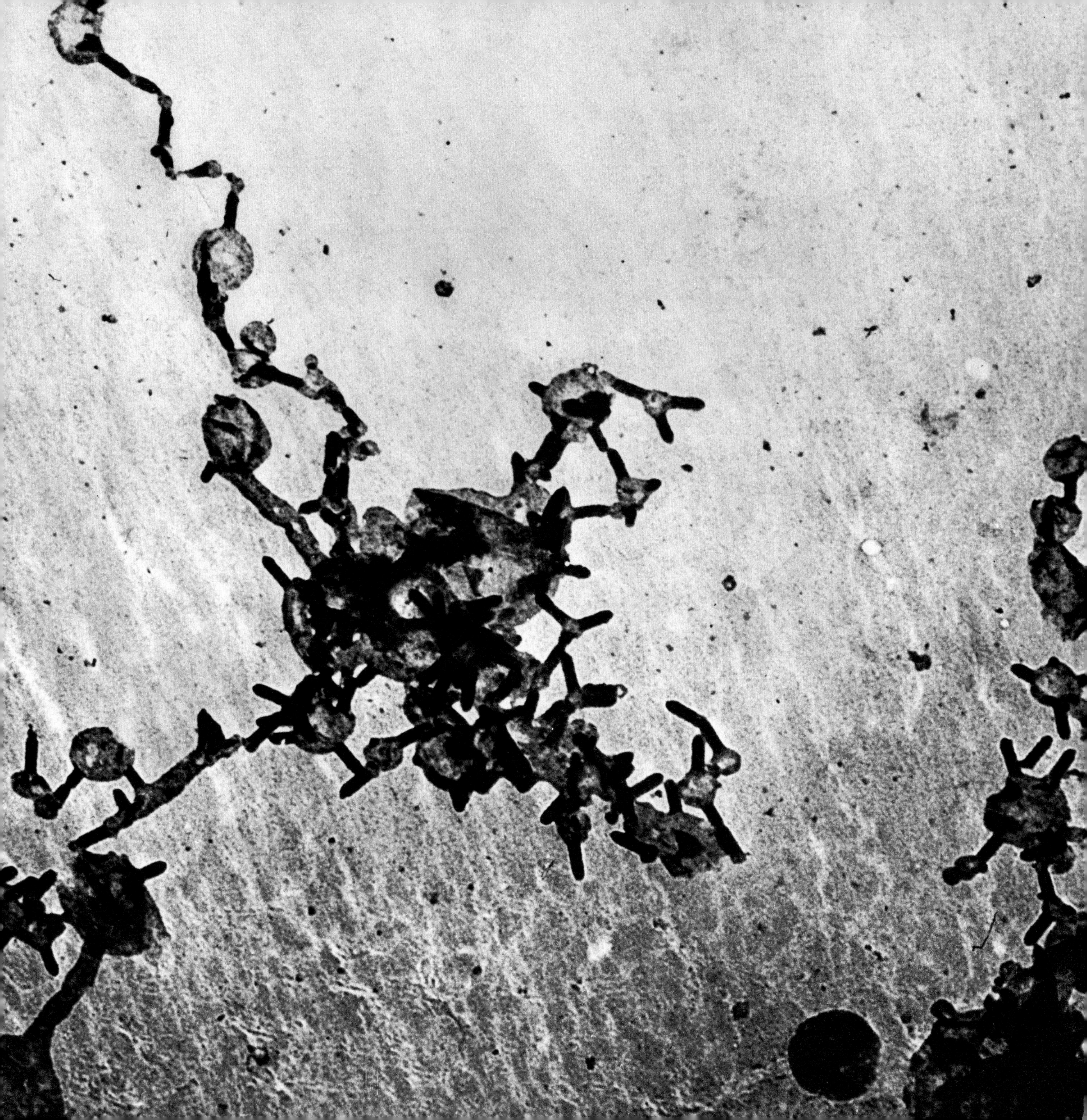

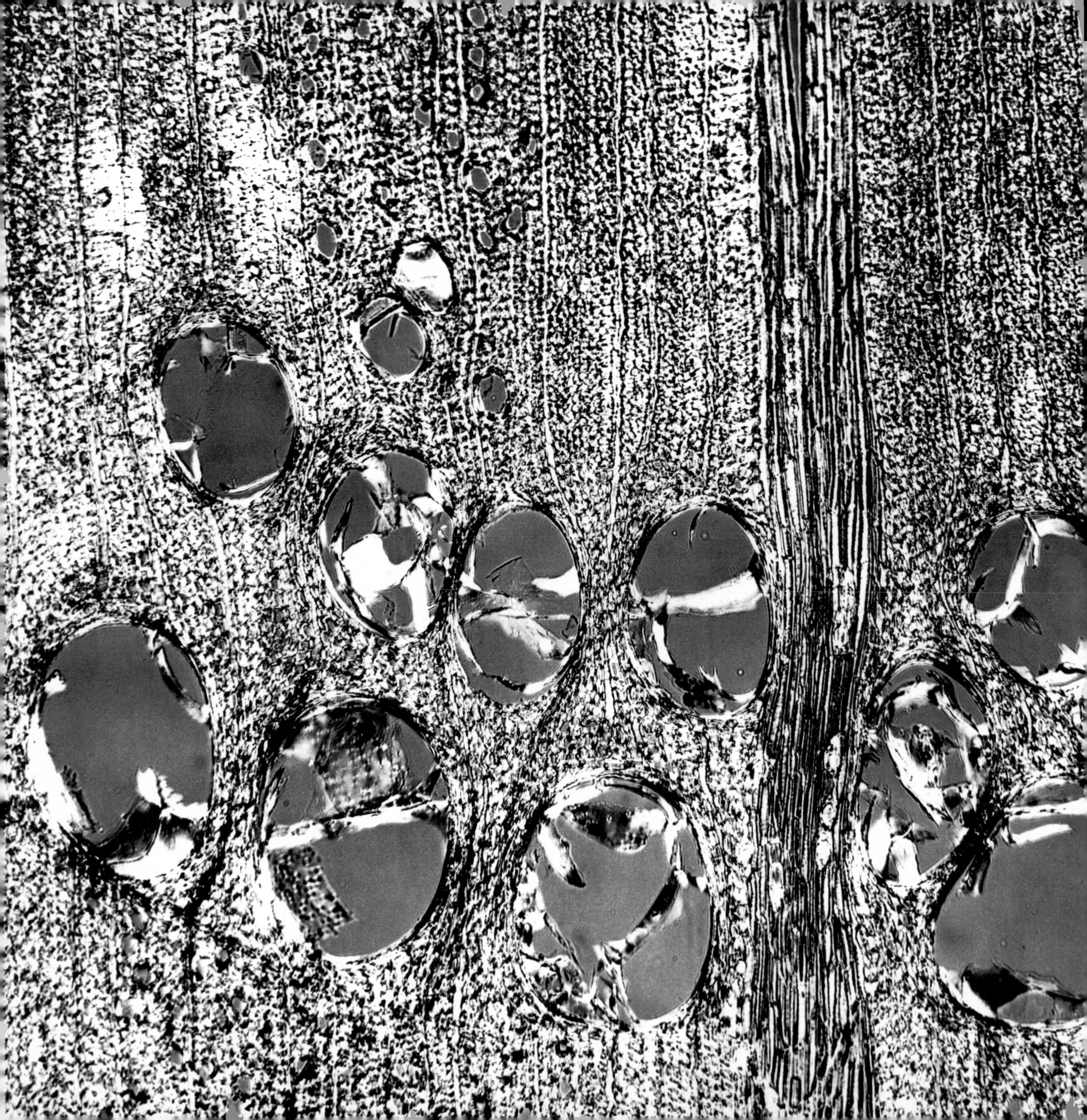

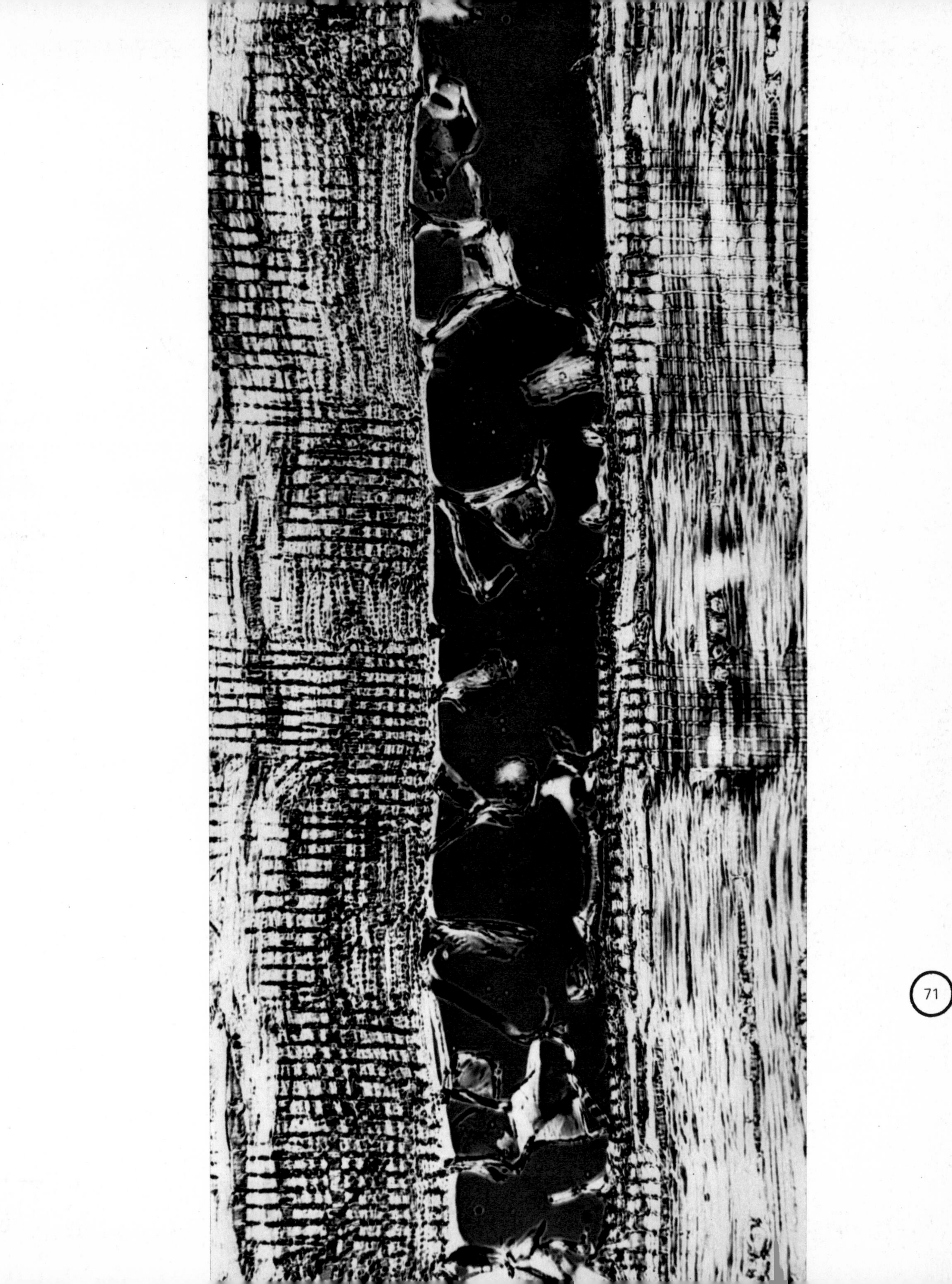

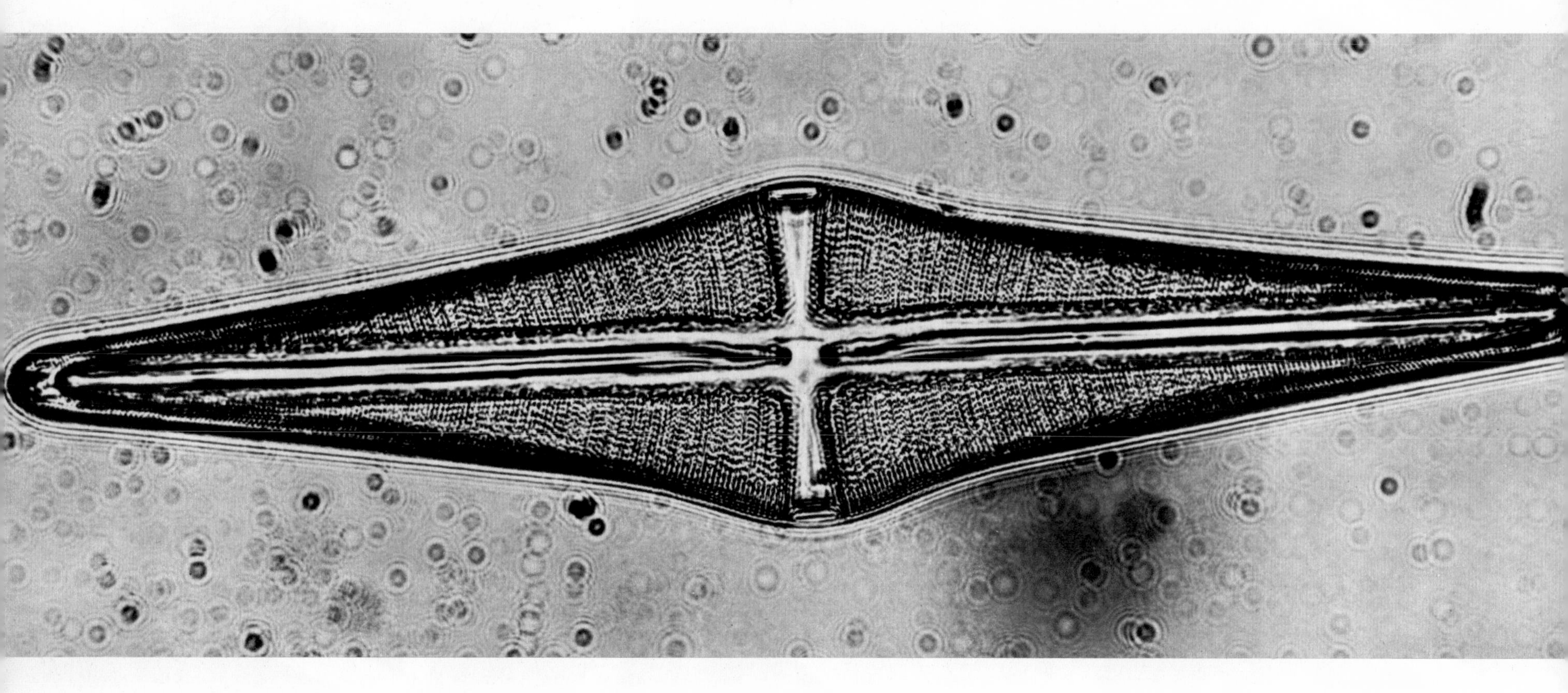

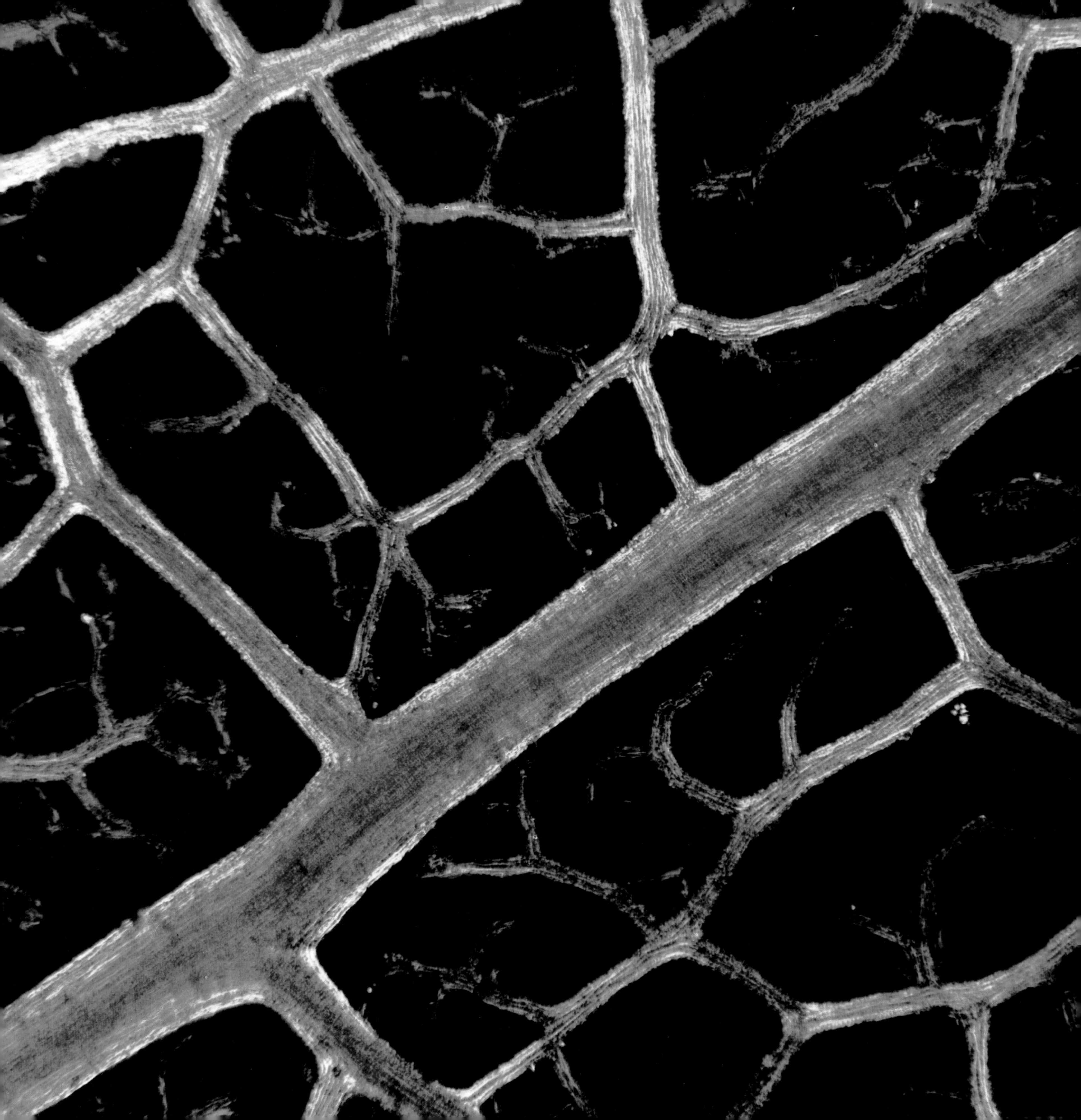

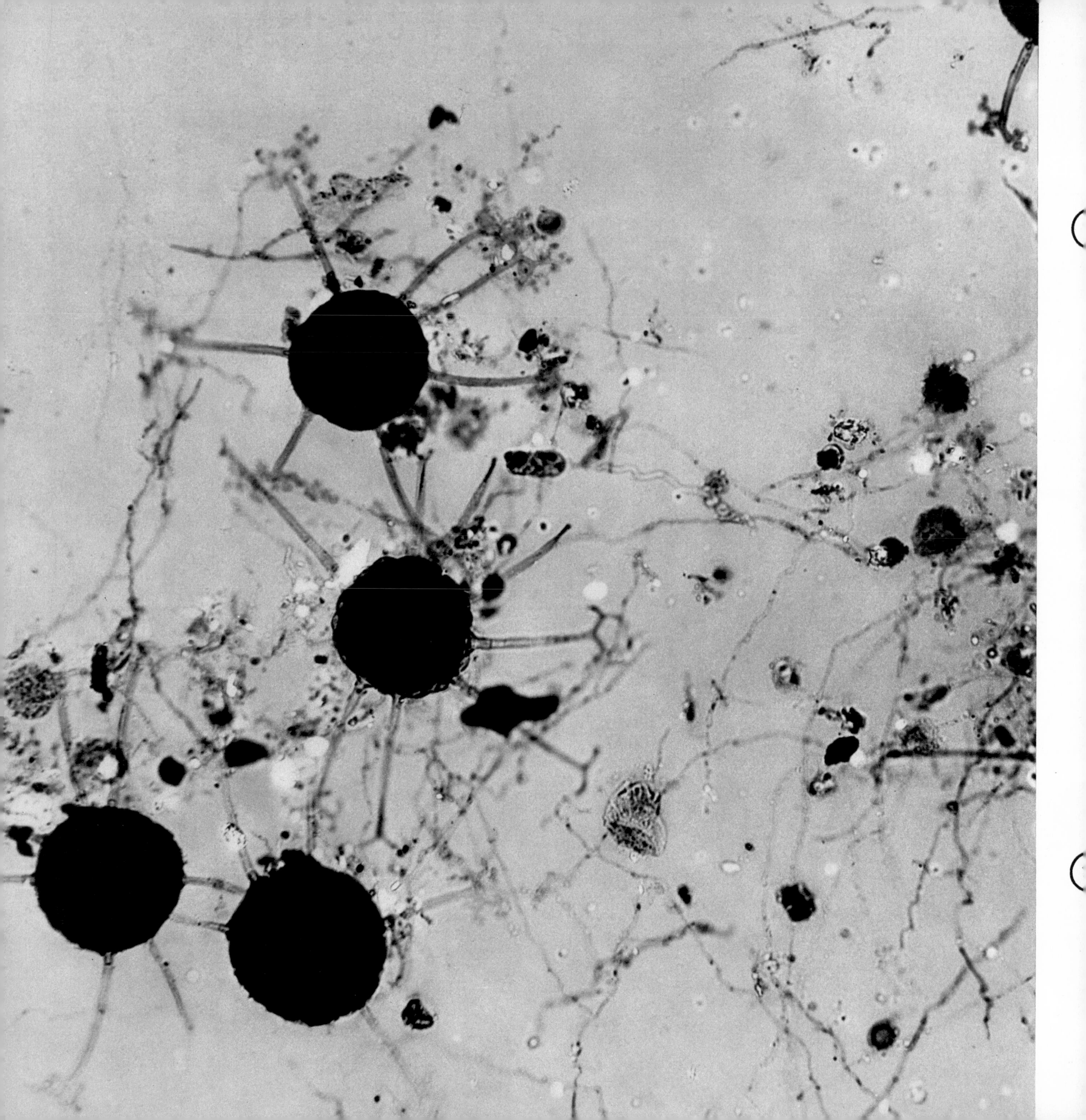

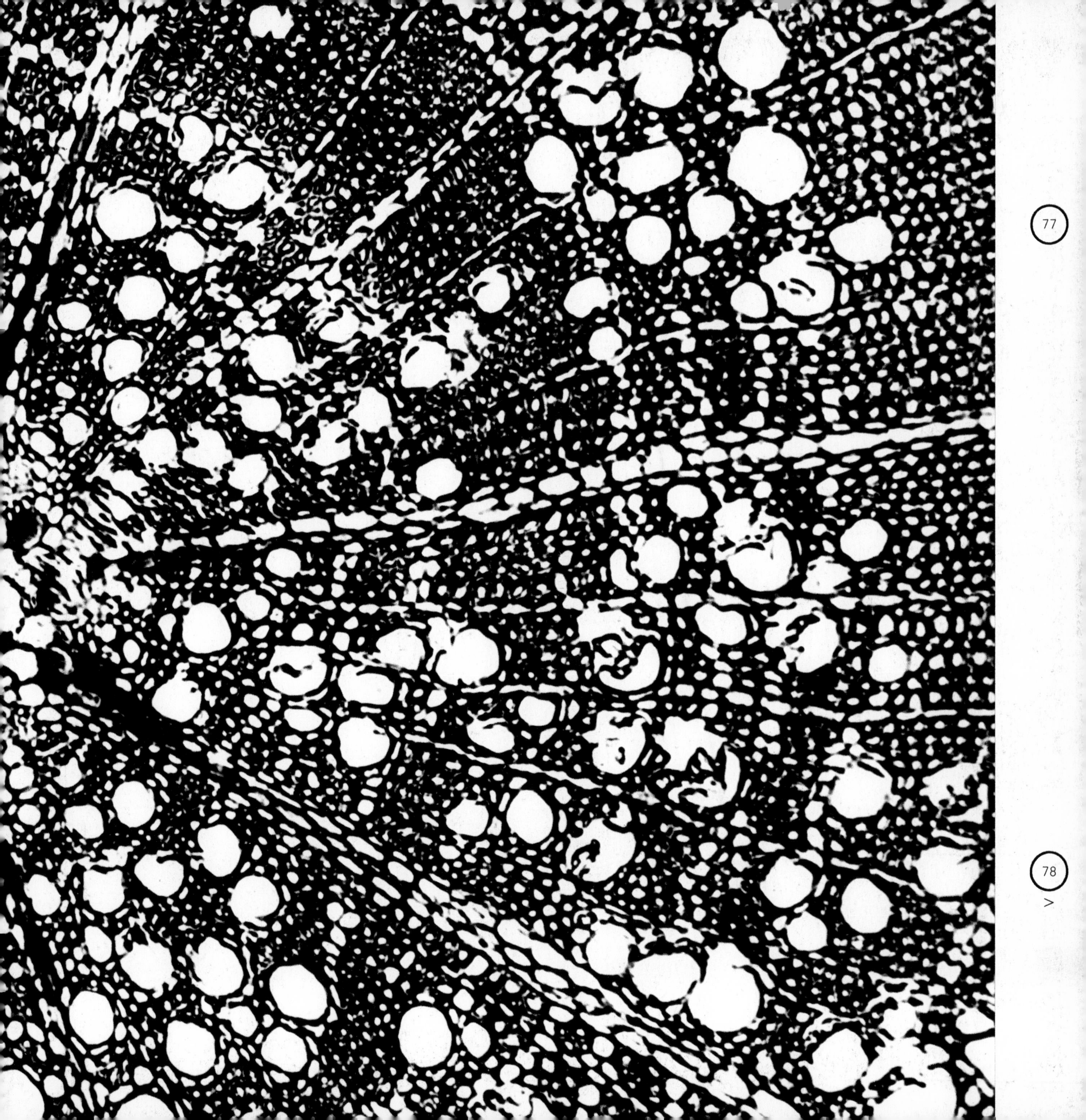

77
78

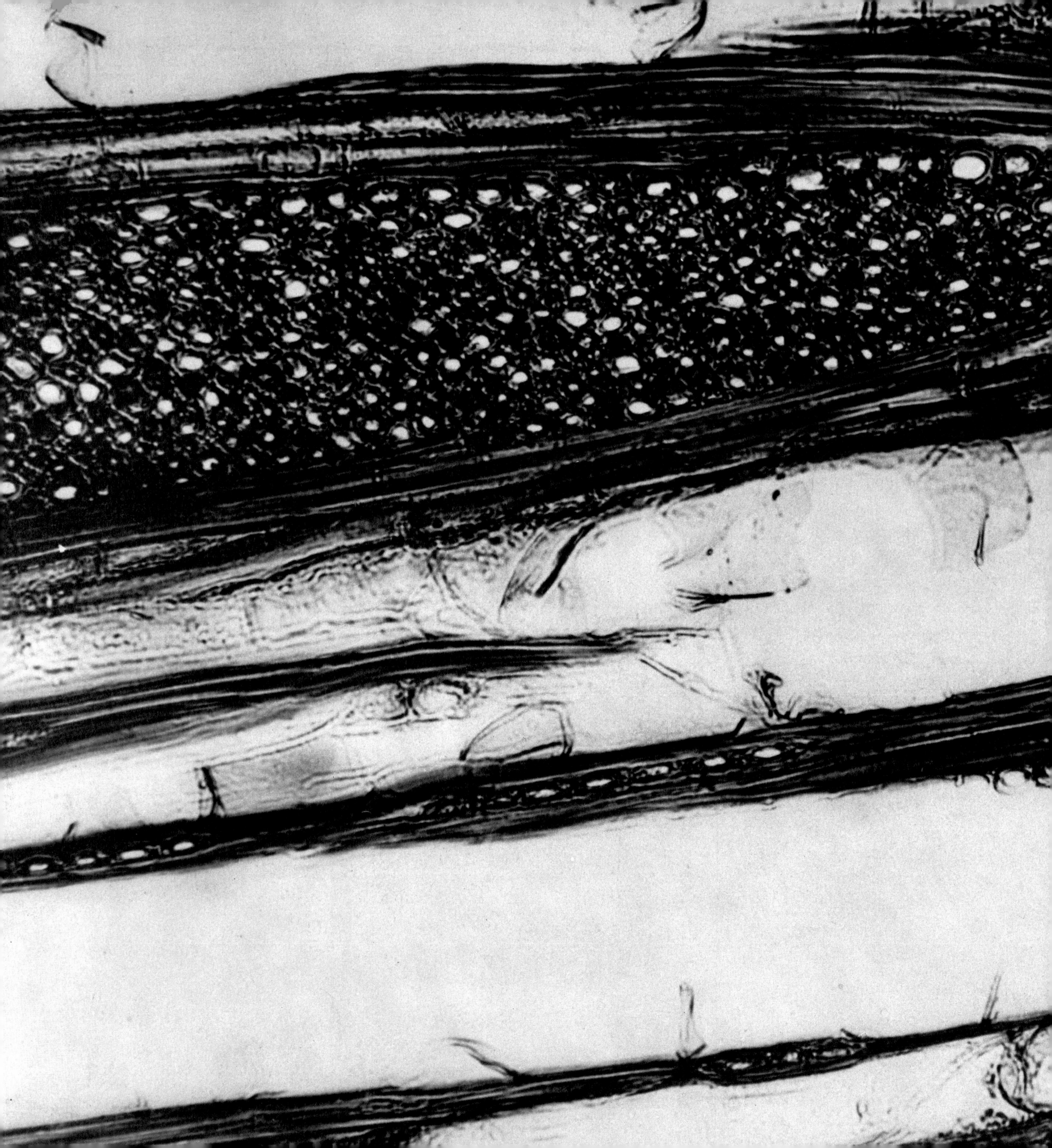

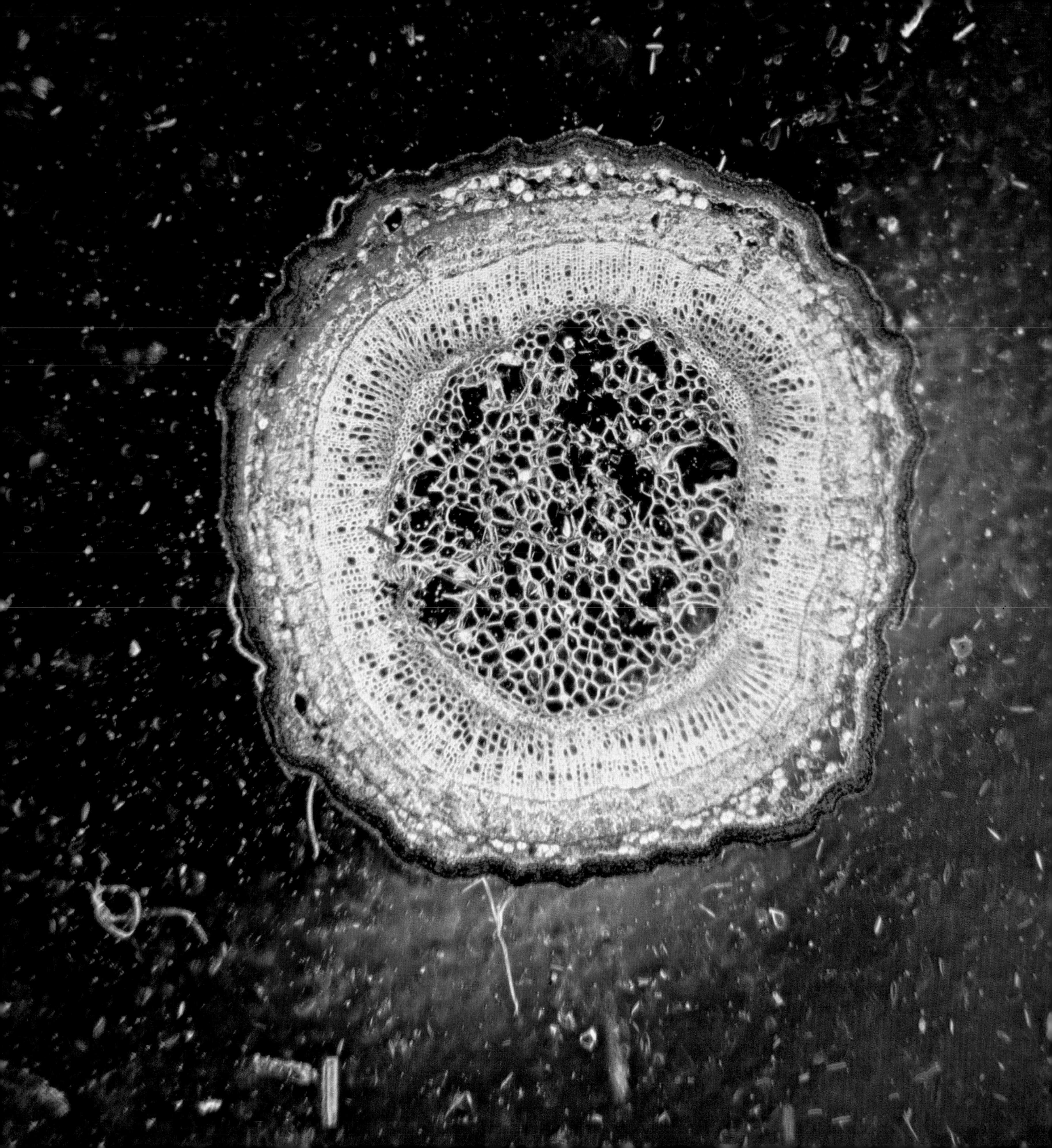

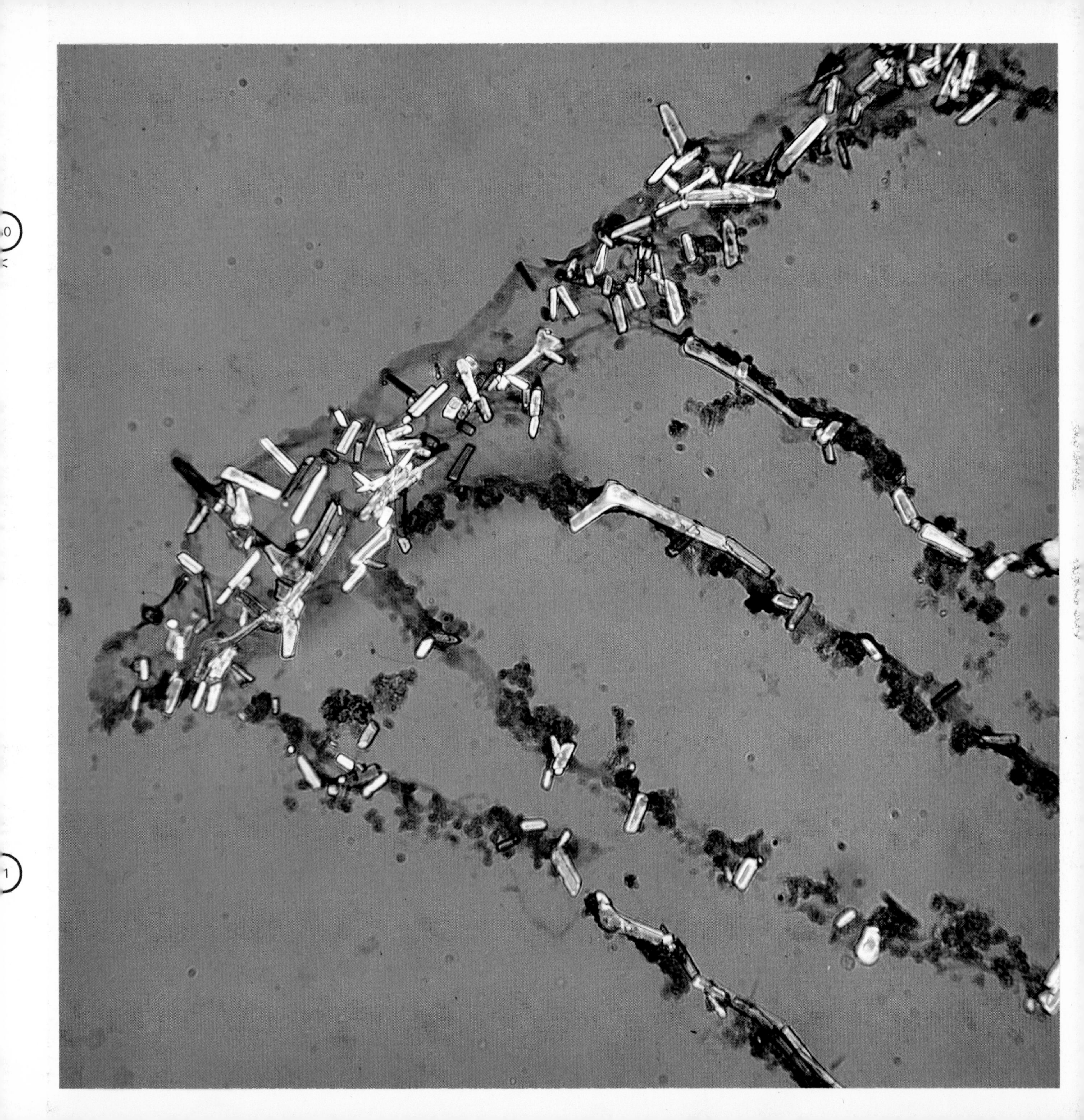

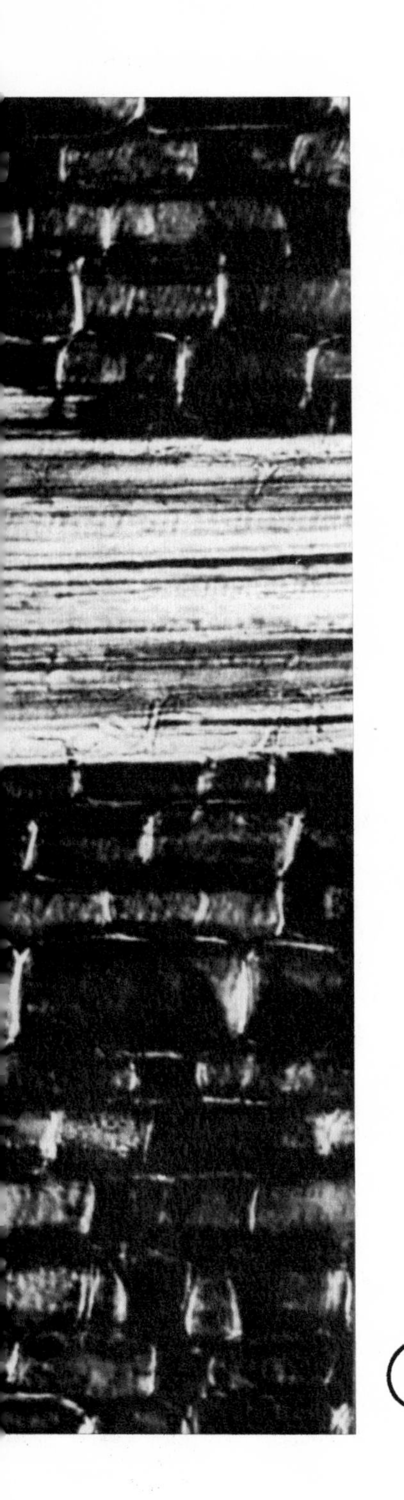

82

83

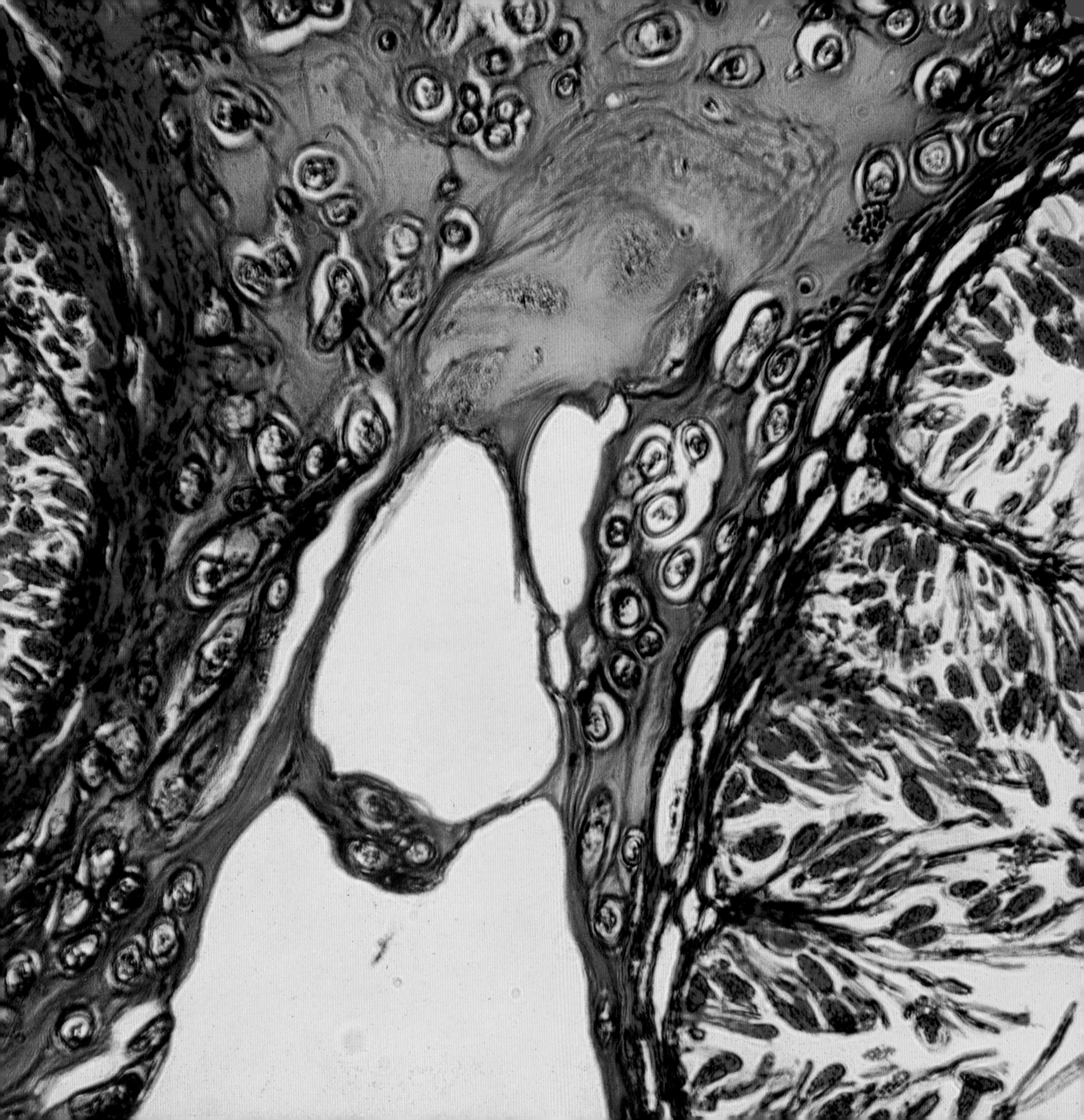

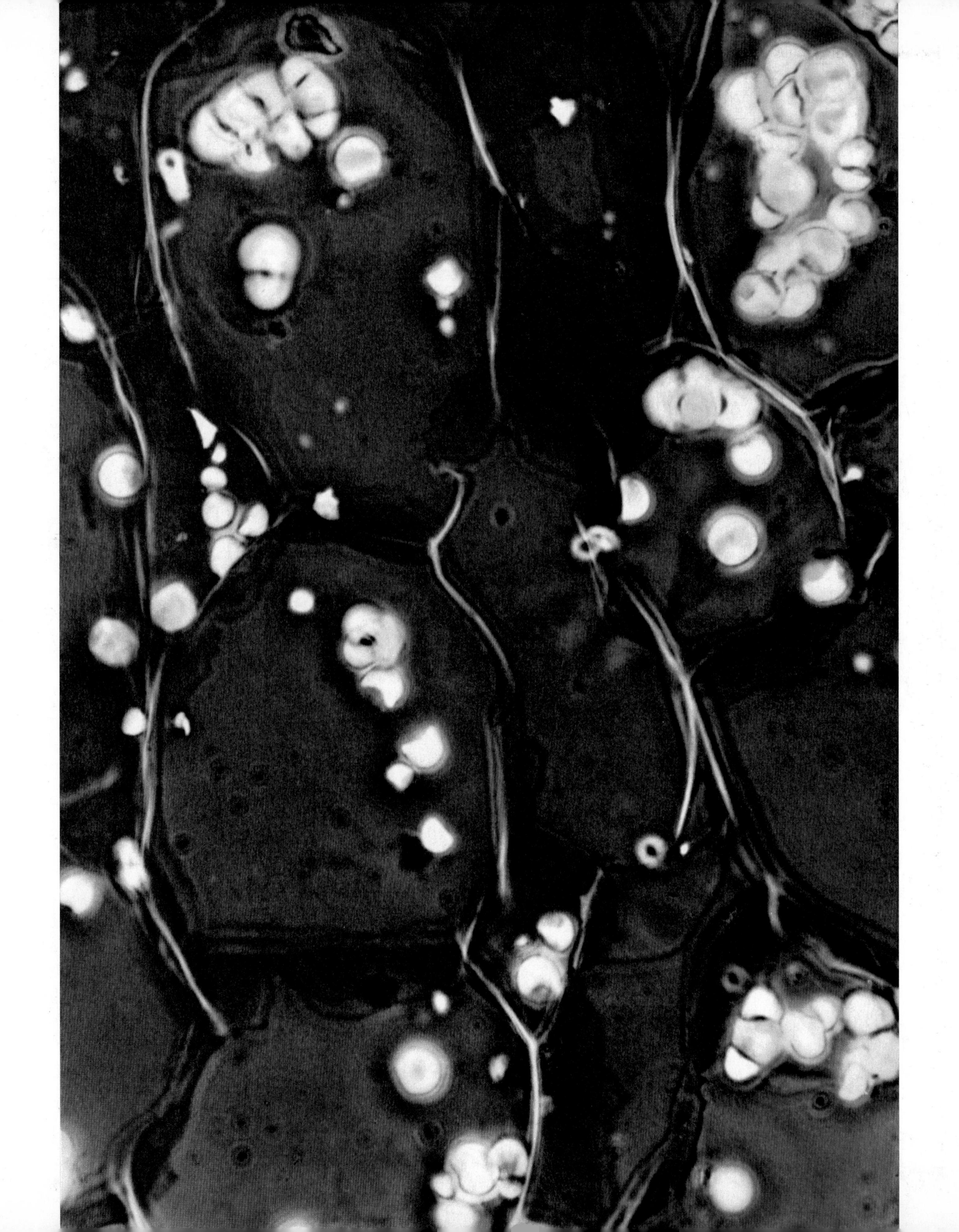

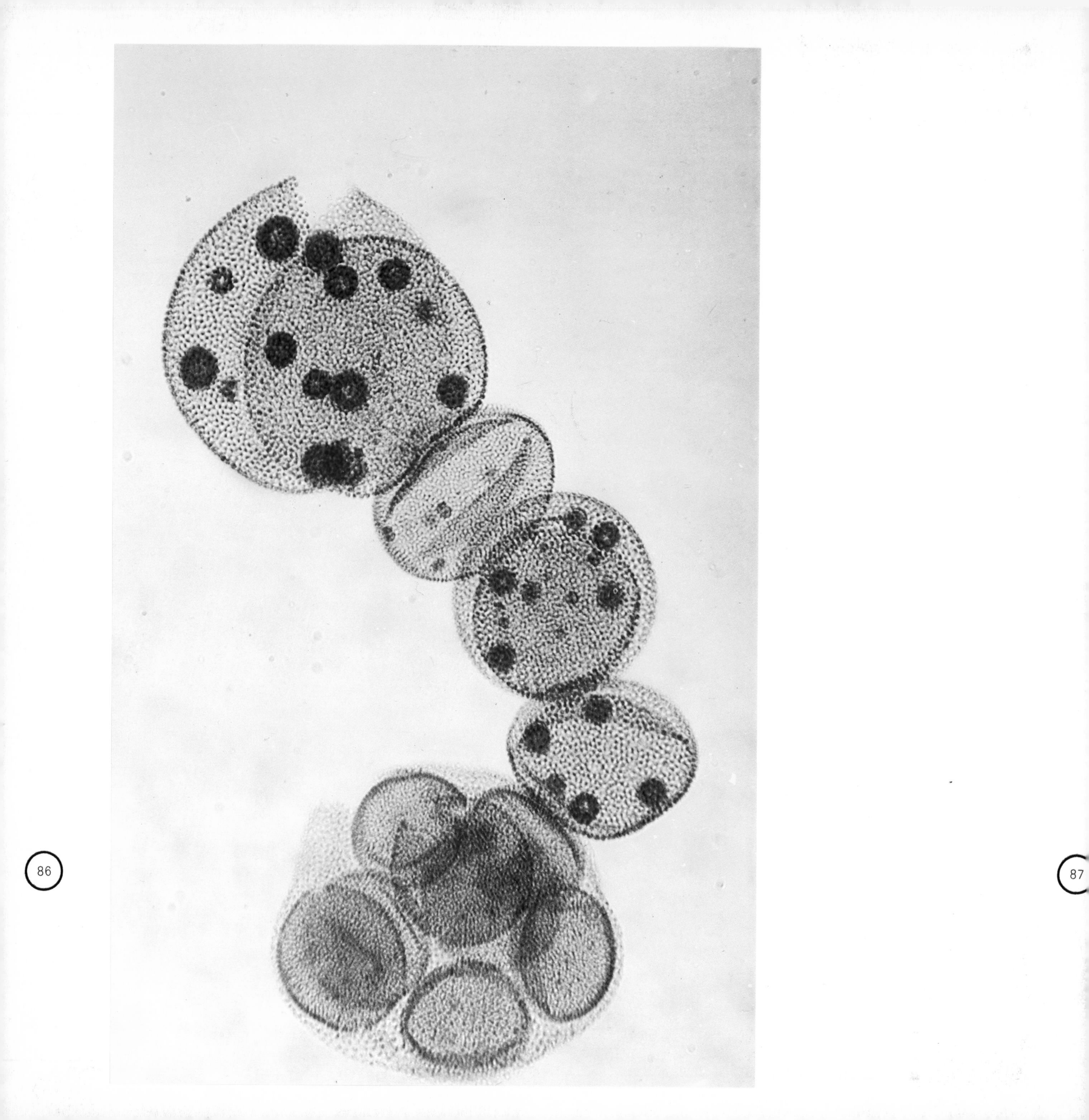

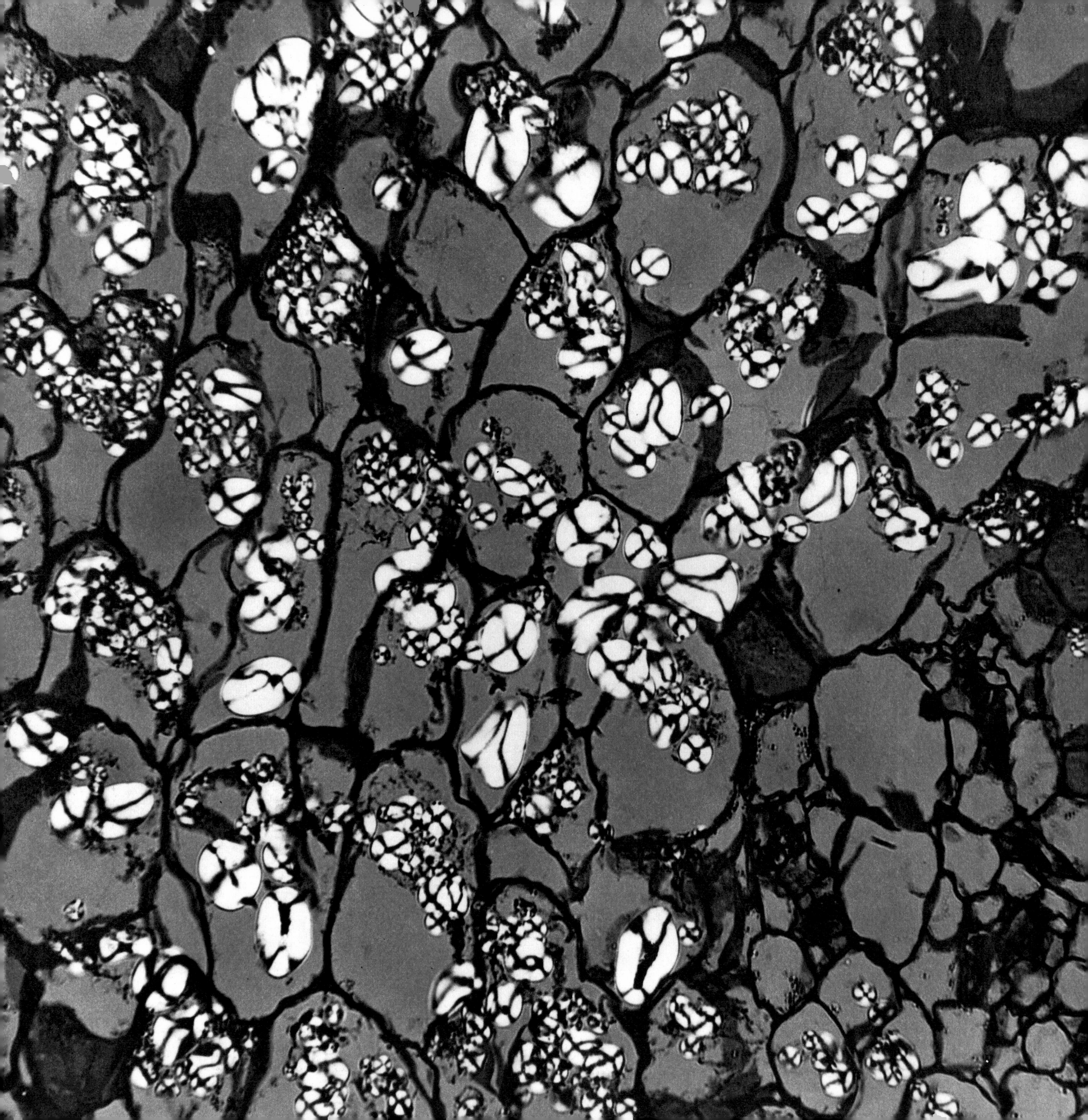

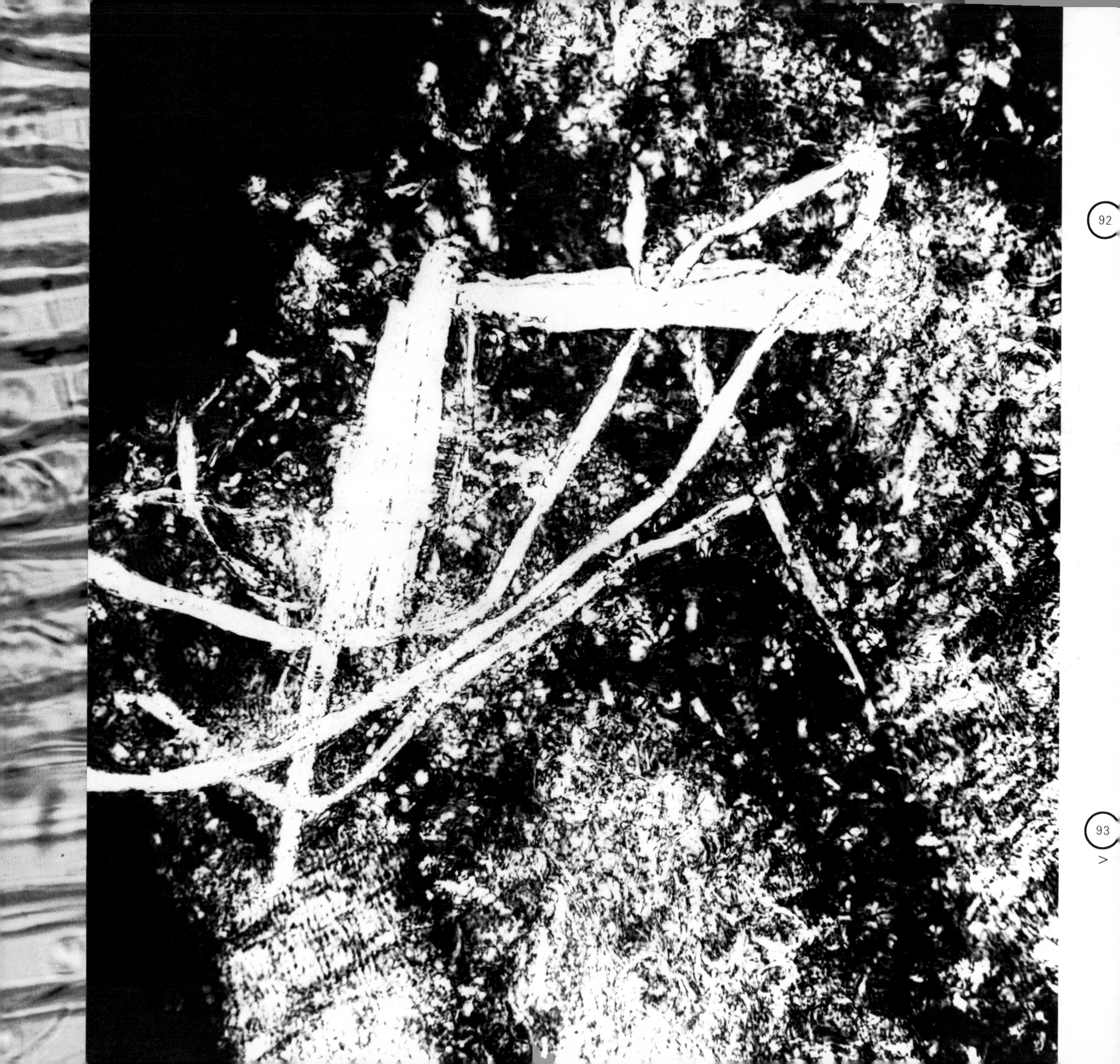

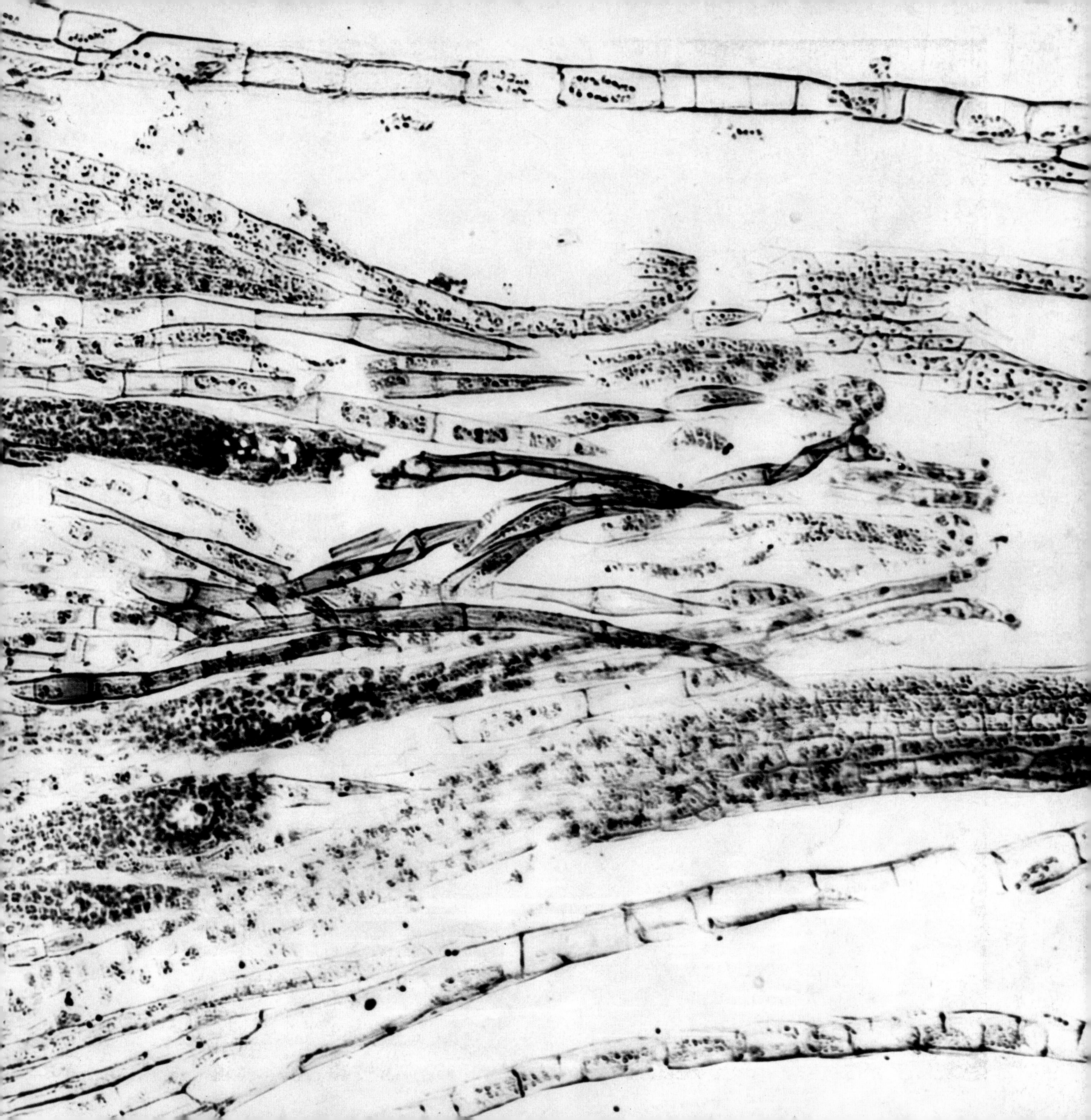

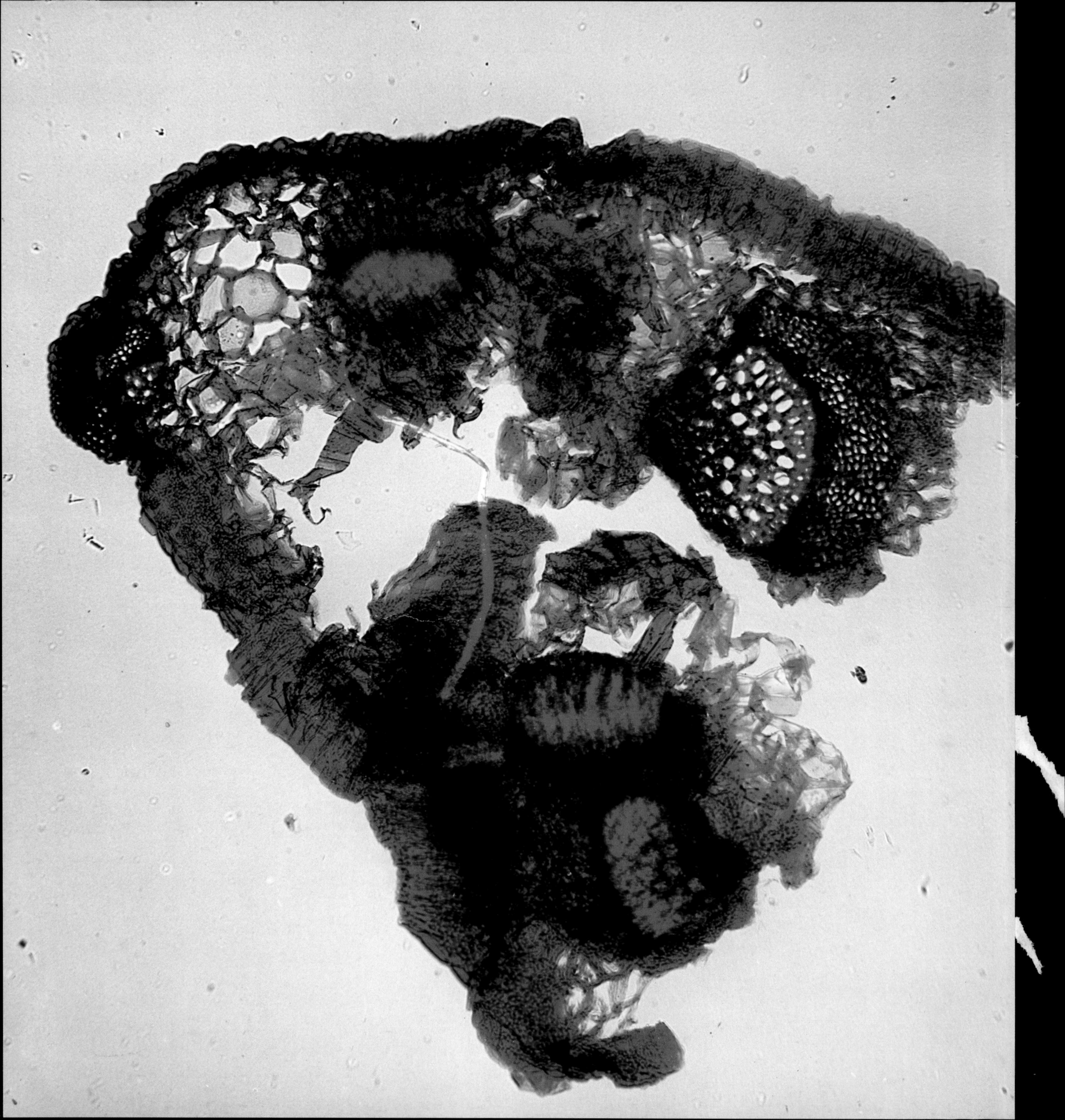

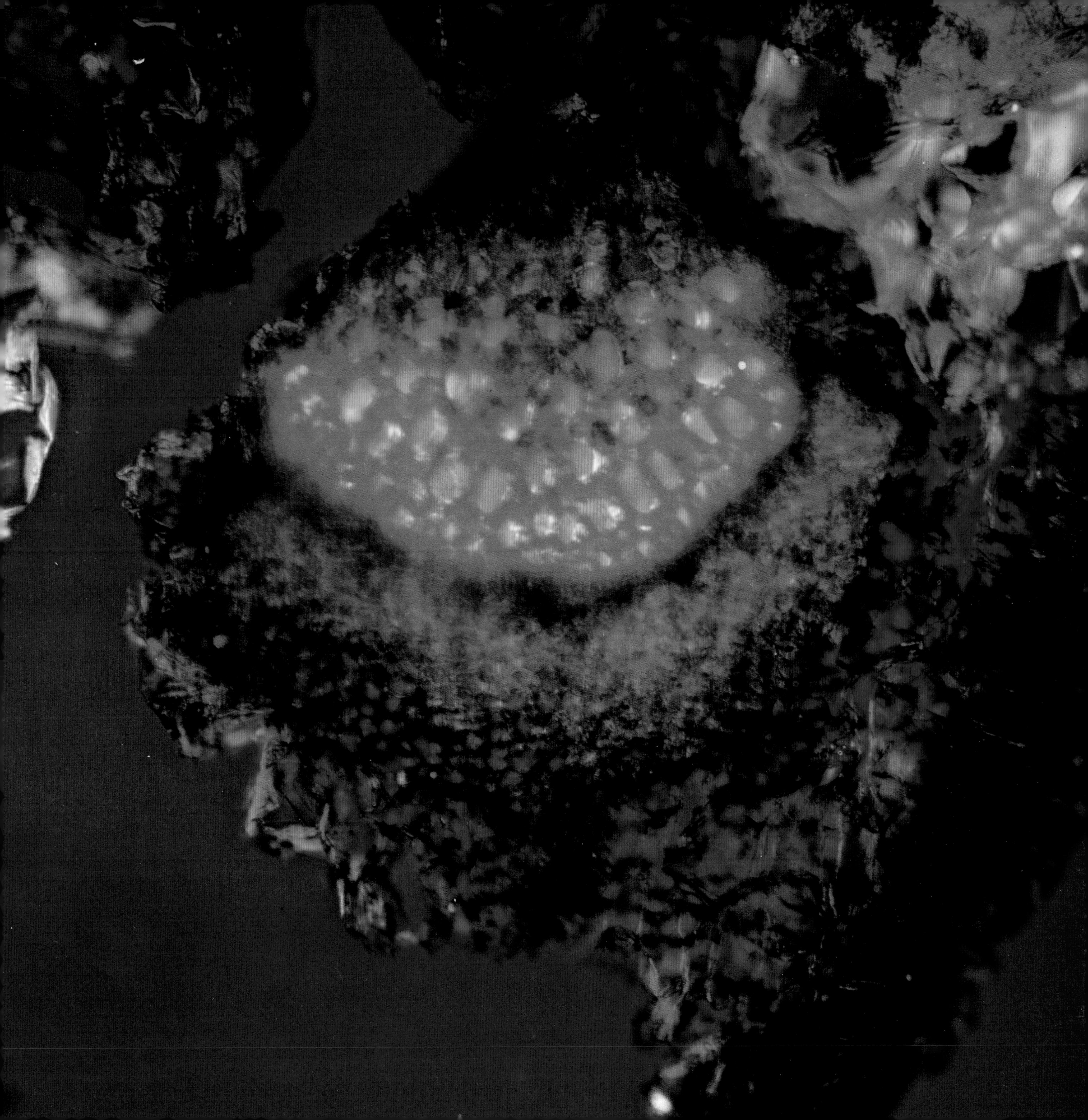

160. Protozoan: *Foraminiferidan*
1,900x

161. Roundworm: *Ascaris*
600x
Cellular cleavage showing mitosis

162, 163. Urethra
750x

163. Protozoan: *Amoeba proteus*
790x

164. Developing tooth
260x

165. Monkey brain
175x

166. Sponge spicules
300x Polarized light

167, 168. Turtle tongue
880x

06. Esophagus
250x

07. Protozoa: *Tetrahymena pyriformis*
750x

08. Cerebellum
790x

09. Umbilical cord
200x

11. Angora rabbit hair
1,100x Polarized light

12. *Hydra* nematocysts
1,500x

13. Protozoan: *Radiolarian*
2,300x

14. Mouse embryo
300x

15. Cow hair
600x Polarized light

16. Protozoa: *Radiolaria* amidst bacteria
1,100x

17. Protozoa: *Paramecia*
1,400x Polarized light

18. External ear
300x

18. Protozoan: fossil *Radiolarian*
800x

19. Protozoan: *Foraminiferidan*
1,900x Polarized light

21. Pituitary
1,150x

22. Protozoan: *Radiolarian*
750x

23. Spermatic cord
235x

24. Protozoa: *Opercularia* colony
580x

25. Mouth parts of male mosquito
175x

26. Aphid
100x

27. Butterfly wing
1,050x Polarized light

28. Butterfly wing
140x Dark field

29. Butterfly wing
1,260x

30. Amphibian ovary: *taricha*
180x

31. Rabbit esophagus

132, 133. Epididymis
460x

134. Lip
1,200x

135. Protozoan: *Stentor*
1,875x

136. Bladder
280x

137. Developing foot of embryo
250x

138. Protozoan: *Radiolarian*
2,000x

139. Shad ovary
400x

right: 140. Flatworm: planaria
580x

left: 140. Flatworm: planaria
230x

141. Frog testes
600x

142, 143. Bat hair
460x Polarized light

144. Fish scale: ctenoid
200x

145. Internal ear of guinea pig: organ of Corti
250x

146. Sciatic nerve
250x

147. Vagina
900x

148. Ureter
800x

149, 150. Radule in digestive system of snail
570x

151. Fish scale: placoid
225x

152. Pigeon esophagus
800x

153. Chinese liver fluke *(Clonorchis sinensis)*
190x

154, 155. Sea anemone: *Metridium*
460x

156. Hydra
210x

157. Pylorus of stomach
300x

158. Trout stomach
700x

159. Protozoan: *Foraminiferidan*

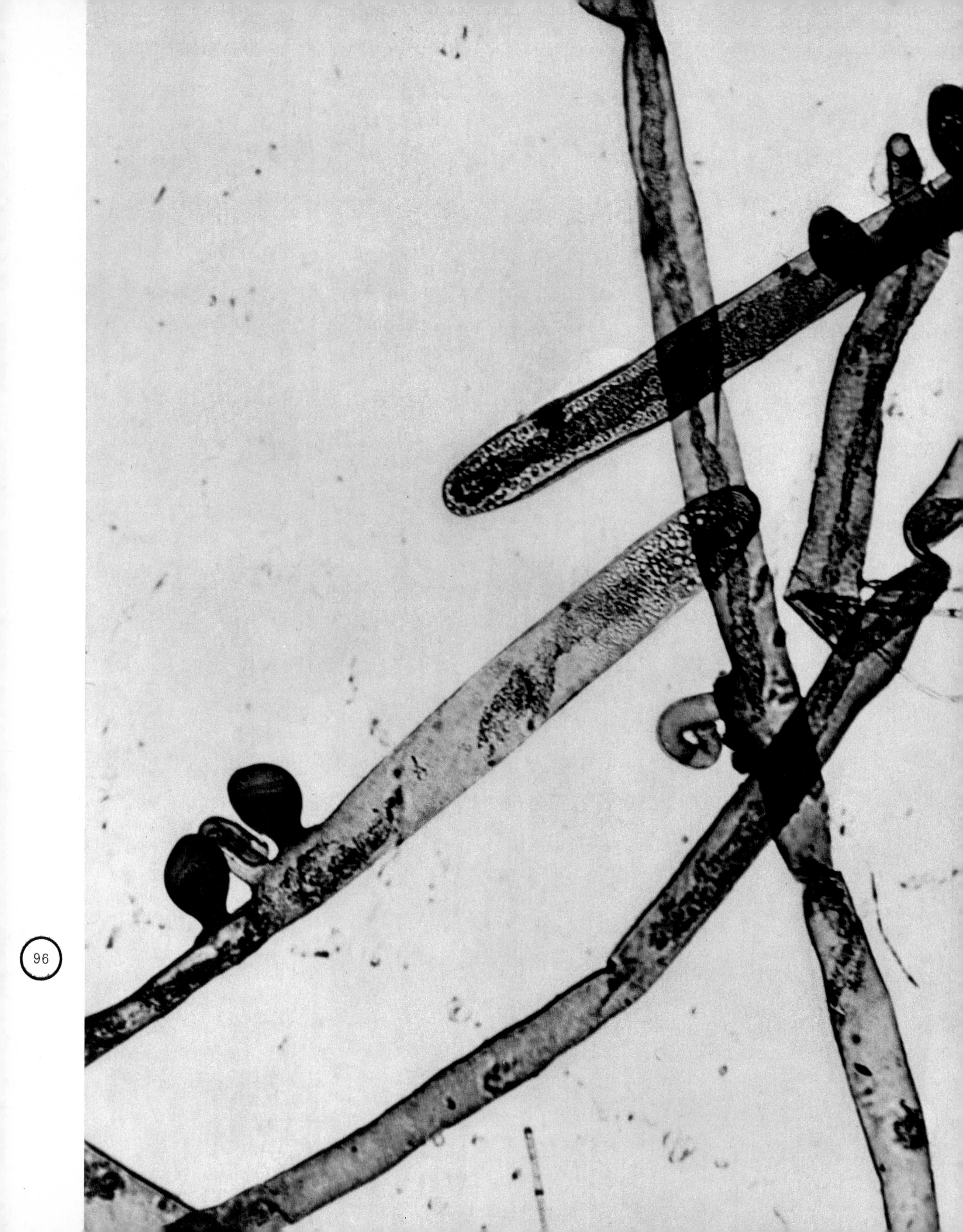

ANIMAL

Keep fold-out (at left) open for captions to the following section

Many people are surprised to learn that mammals are not the only members of the animal kingdom. Reptiles, birds, fish, shellfish, insects, worms, sponges, and even creatures like the amoeba and paramecium are, in fact, all true animals.

The basic unit of structure in all living matter is the cell. It is essentially the same in animals and plants—both possess tissues that are identical in chemical composition. The differences between animals and plants have already been outlined in terms of form, structure, and life functions. The relationship of plants to animals is one of interdependence: green plants provide oxygen and are the ultimate source of all food for animals; animals provide carbon dioxide and minerals essential for the growth of plants.

In visual terms, there is as much variety in animal tissue as in vegetable tissue. As for differentiating the two categories, there is no systematic way of doing so—some of the animal images greatly resemble those of plants. The main observation to be made is that all these images are beautiful, all excite the imagination, and all reveal the extraordinary reality of inner space.

One-celled animals

It was approximately three hundred years ago that Leeuwenhoek, using a primitive microscope and a crude illumination source, perceived to his great amazement unicellular organisms swimming about in a drop of water on a slide. To these objects he assigned the name "animalcules," and he proceeded to describe with great clarity the forms and some of the habits of these creatures. His letter to the Royal Society announcing his discovery is fascinating. The following is an excerpt:

In the year 1675, *about half-way through September (being busy with studying air, when I had much compressed it by means of water), I discovered living creatures in rain, which had stood but a few days in a new tub, that was painted blue within. . . .*

Of the first sort that I discovered in the said water, I saw, after divers observations, that the bodies consisted of 5, 6, 7, *or* 8 *very clear globules . . . . When these animalcules bestirred 'emselves, they sometimes stuck out two little horns, which were continually moved, after the fashion of a horse's ears. The part between these little horns was flat, their body else being roundish, save only that it ran somewhat to a point at the hind end; at which pointed end it had a tail, near four times as long as the whole body, and looking as thick, when viewed through my microscope, as a spider's web. At the end of this tail there was a pellet . . . . These animals were the most wretched creatures that I have ever seen; for when, with the pellet, they did but hit on any particles or little filaments (of which there are many in water, especially if it hath but stood some days), they stuck intangled in them; and then pulled their body out into an oval, and did struggle, by strongly stretching themselves, to get their tail loose; whereby their whole body then sprang back towards the pellet of tail, and their tails then coiled up serpent-wise, after the fashion of a copper or iron wire that, having been wound close about a round stick, and then taken off, kept all its windings.*

This is an account of the infusorian *Vorticella*. In the same graphic way, Leeuwenhoek described a number of protozoa, including *Foraminifera*, the ciliates *Coleps* and *Cothurnia*, and the flagellates *Anthophysa* and *Volvox*.

Leeuwenhoek's findings stirred up great controversy. The tiny size, massive numbers, and astonishing capacity for multiplication of these microscopic creatures made it seem plausible in the minds of some scientists that they could be accounted for only by the theory of spontaneous generation—life originating from nonliving matter. In 1718, Louis Jablot exploded this theory by demonstrating that sealed jars of boiled water containing "animalcules" were devoid of life. Around the middle of the eighteenth century, owing to the work of such pioneers as John Hill, O. F. Mueller, C. G. Ehrenberg, and F. Dujardin, a wealth of data was accumulated on the morphology and physiology of protozoa. It became apparent that one could study the processes of life and the structure of living matter more readily in these one-celled creatures than in more highly specialized organisms. Out of these observations emerged cytology, the science that deals with the structure and function of cells.

As the microscope improved, increasing numbers of primitive animals were identified, each protozoan a complete, independent organism. In the course of evolution, it was hypothesized, the joining of some of these units in colonies and the gradual specialization of cell groupings created a differentiation of tissues characteristic of multicellular organisms—the *Metazoa.*

Protozoa are divided into four classes: (1) *Mastigophora,* or *Flagellata,* which propel themselves with whip-like tails called *flagella* (example: *Euglena*); (2) *Sarcodina,* or *Rhizopoda,* which move about by flexible arm-like extensions, or *pseudopods* (example: *Amoeba*); (3) *Ciliophora,* or *Ciliata,* which are covered with hair-like projections that propel the animal (example: *Paramecium*); (4) *Sporozoa,* which are carried about by other unicellular creatures in the form of a cyst, or enclosed envelope (example: *Plasmodium*). The first of these classes, *Mastigophora,* is so primitive that it represents, as mentioned in the previous section, a developmental phase somewhere between plant and animal cells. It possesses chloroplasts, the green substances of plants, which enable them to manufacture their own food through the energy of sunlight; but in the absence of light, *Mastigophora* can, like animals, engulf food materials around it. As a consequence, a lively debate goes on between biologists and zoologists as to whether flagellates belong to the animal or plant kingdom. Perhaps it is correct to assign them to both.

Protozoa feed on bacteria (a single *Paramecium* consumes over five million bacteria daily) and on each other (the diet of the *Didinium* consists solely of *Paramecia,* and the *Spathidium* largely subsists on *Colpidia*). They usually function independently, although certain species cling together in colonies, swimming as a group by the combined action of the flagella (for example, *Gonium, Volvox, Pandorina, Sudorina*). Protozoa exist in many shapes, ranging from the spherical *Radiolaria,* to the asymmetrical elongated *Paramecia,* to the variegated forms of *Foraminifera.*

When environmental conditions such as moisture and temperature become unfavorable, many protozoa have the ability to encyst themselves within a thick wall and live in a kind of suspended animation. After months, and even years, if conditions become favorable, these *spores,* or *zygotes,* are able to resume their active state. Reproduction is extremely rapid; it is possible for a few active protozoa to generate millions of descendants in a matter of hours.

Protozoa abound in ponds, lakes, seas, wet soil, and the bodies of higher plants and animals, which act as their hosts. There are more kinds of protozoa than any other creatures in the animal kingdom, thirty thousand species having been identified. Among the most beautiful protozoa, from a photographic standpoint, are those that secrete skeletal structures in the form of shells. Of particular interest are the *Radiolaria,* which have internal lattices of silica, and the *Foraminifera,* whose endoskeletons are made of lime. These shells are perforated with pores through which protoplasmic projections (*rhizopods*) pass as the organism captures food and propels itself through its environment. The skeletons of these creatures make up a good part of the ooze that covers the floor of the ocean, and, with sponge spicules and the remains of diatoms from past geological epochs, constitute the massive chalk rocks found on land and on the ocean floor.

Many-celled animals

From the standpoint of the photomicrographer, there is surprisingly little difference between animal and plant tissues. The cells of each are essentially composed of the same materials. Indeed, as has been pointed out, some of the unicellular varieties of both branches are so much alike that there is some doubt as to which classification they fit best. Both plant and animal cells go through similar vital processes. They require food for their energy, growth, and reproductive needs. Their protoplasm contains proteins, carbohydrates, fats, inorganic salts, and water. Vegetable cells contain intricate chemical substances (the chloroplasts) that manufacture complex organic chemicals from simple materials found in the soil and air. Animal cells are not so self-sufficient, requiring carbohydrates and proteins, which they incorporate parasitically by feeding on plants or other animals.

The body of multicellular animals is a system of tubular or spherical cell masses constructed around ingestive, digestive, and eliminative food tracts. Plants require sunlight for the fueling of their chemical machinery; thus plant tissues spread themselves out over large exterior areas to take advantage of every ray of sunlight. But animal tissues are more concentrated. Compared with plants, animal structures adjust more readily to the demands of their environment; and, in the course of evolution, they have developed a great diversity of shapes and organs. The basic life functions are nevertheless common to both plants and animals. Oxygen, water, and food must be acquired, and such vital activities as respiration, circulation, digestion, secretion, and reproduction must be negotiated. The internal systems assigned to these tasks vary with the complexity of the organism.

The only major group of multicellular animals without a digestive cavity are the *sponges,* which dwell on ocean floors. A stiff skeleton supports a porous body that soaks in microscopic food particles..The outer surface is covered with sensitive contractile cells controlling the input and outgo channels as well as the overall movement of the sponge. There are five thousand species of sponges, some as tall as man.

The *coelenterates*, represented by ten thousand species, are beautifully structured, symmetrical organisms with tissues arranged around a hollow container, either vase-shaped or bowl-shaped. Within the container, digestive enzymes are released to process the food, which, in turn, is absorbed by the living cells. Inedible remains are disgorged. Tentacles contain specialized cells with thread capsules that are shot out to poison, paralyze, and rope in their prey. There are three classes of coelenterates: the *Hydrozoa*, which possess a simple nerve network (the Portuguese man-of-war is a colony of *Hydrozoa* attached to a gas-filled sac); the *Scyphozoa*, or jellyfish, which has true muscle cells, special sensory cells, and two sense organs; and the *Anthozoa*, which includes sea anemones and corals.

Climbing the evolutionary scale, we come to the fifteen thousand species of *flatworms*, the first organisms with a distinctive nervous system and the beginnings of a brain. Next we come to the *annelids*, elongated and segmented invertebrates, having a one-way digestive tract, five heart pairs, which are part of a circulatory system, and an elaborate nervous system. Annelids consist of seven thousand species, among which are marine worms, earthworms, leeches, lampshells, moss animals, and many other wormlike animals.

The *mollusks*—one hundred thousand species—are divided into five classes and include the following familiar animal creatures: clams, oysters, scallops, mussels, snails, slugs, whelks, squid, octopus, and nautilus. The *echinoderms*, fifty-seven hundred species, are radially symmetrical marine animals, equipped with a water-vascular system that acts as a hydraulic device for suction, movement, and attack. In this category are sea urchins, sand dollars, brittle stars, serpent stars, sea lilies, feather stars, and sea cucumbers. By far, the joint-footed *arthropods* are the most complex of the invertebrates. They constitute 80 percent of the animal kingdom, numbering no less than one million separate species. To describe the many species of arthropods would require the space of an encyclopedia. They are grouped into seven classes and include the following animals: lobsters, crabs, crayfish, shrimps, horseshoe crabs, spiders, scorpions, mites, beetles, butterflies, bees, ants, fleas, lice, bugs, centipedes, and millipedes. The adaptation of the organs of these creatures to their environments is a fascinating commentary on the marvels of evolution. The social life of these creatures is no less astonishing, some communities having rigid caste systems going back one hundred million years.

The *chordates*, highest phylum on the evolutionary scale, consist of about forty-five thousand species. They range from wormlike organisms, to fishlike marine animals, to the vertebrates, whose highest evolutionary product is, of course, man.

The animal tissues illustrated are just a handful of the near infinity of choices. The selection is based exclusively on visual variety, in the attempt to show some part of the tableau of extraordinary designs and textures that exist in multicellular organisms.

Kingdom, *Animalia*—Phylum, *Chordata*—Subphylum, *Vertebrata*—Superclass, *Tetrapoda*—Class, *Mammalia*—Order, *Primata*—Family, *Homidae*—Genus, *Homo*—Species, *sapiens*. This is man, the paragon of animals, who has risen from microscopic beginnings to dominate all other living things on earth, and who now is reaching into extraterrestrial domains. The last link on a long chain of genetic variations, surviving as a consequence of natural selection, man still bears evidence in his tissues of his affiliation with primitive animals and even with plants. He preys on plants and other animals, much like a protozoan. He has embryonic gills in his pharynx, just like a proboscis worm. His spinal cord is encased in a vertebral column and his skull encloses his brain, like any bony fish. He is a land vertebrate, as is a salamander. He has skin with hair, a respiratory tract with diaphragm, a constant body temperature, red corpuscles lacking nuclei, and milk-producing glands, much like a simple opossum. He has distinctive fingers and flat nails, like a lemur. His color vision, flat face, and ability to stand on two legs are characteristics common to any ape.

What really distinguishes man from other creatures is his gigantic brain, his speech facilities, his capacity for conceptual thought, his long childhood through which cultural indoctrination is implemented, and his highly organized social system, which includes government, law, education, and religion.

From scientific studies of all aspects of life processes and cellular development, it is clear that human beings emerged from the same cradle as all other living things. Indeed, a good many links in the long chain from unicellular organisms to man continue to live on earth in forms similar to those of their primitive forebears. Many such organisms are now extinct and are found only as fossils encased in rock or soil. But combining all the clues now available, we can construct some reasonable hypotheses about how living things evolved into their present form. We have been able to trace a direct descent from fish to amphibians, to reptiles, to birds, and finally to mammals—of which man is the highest form.

There are approximately thirty trillion cells in the human body serving such diverse functions as metabolism, growth, irritability, and reproduction. As in all organisms, human cells are structured for different functions and fall into many categories, relating to the following bodily systems: the *integumentary* system (with structures such as skin, hair, and nails), the *skeletal* system, the *muscular* system, the *digestive* system, the *circulatory* system, the *respiratory* system, the *excretory* system, the *endocrine* system, the *nervous* system, the *sensory* system, and the *reproductive* system. In each of these categories, the cells have different functional and visual characteristics.

In the last paragraph of *The Descent of Man*, Charles Darwin writes:

Man may be excused for feeling some pride at having risen, though not through his own exertions, to the very summit of the organic scale; and the fact of his having thus arisen, instead of having been aboriginally placed there, may give him hope for a still higher destiny in the distant future. But we are not here concerned with hopes and fears, only with the truth as far as our reason permits us to discover it; and I have given the evidence to the best of my ability. We must, however, acknowledge, as it seems to me, that man with all his noble qualities, with sympathy which feels for the most debased, with benevolence which extends not only to other men but to the humblest living creatures, with his god-like intellect which has penetrated into the movements and constitution of the solar system—with all these exalted powers—man still bears in his bodily frame the indelible stamp of his lowly origin.

A microscopic examination of the tissues of man will clearly reveal their resemblance to the tissues of lower animals. The cellular groupings are typical samples of human tissue. By no means comprehensive, the selection was made purely on the basis of variety and esthetic merit.

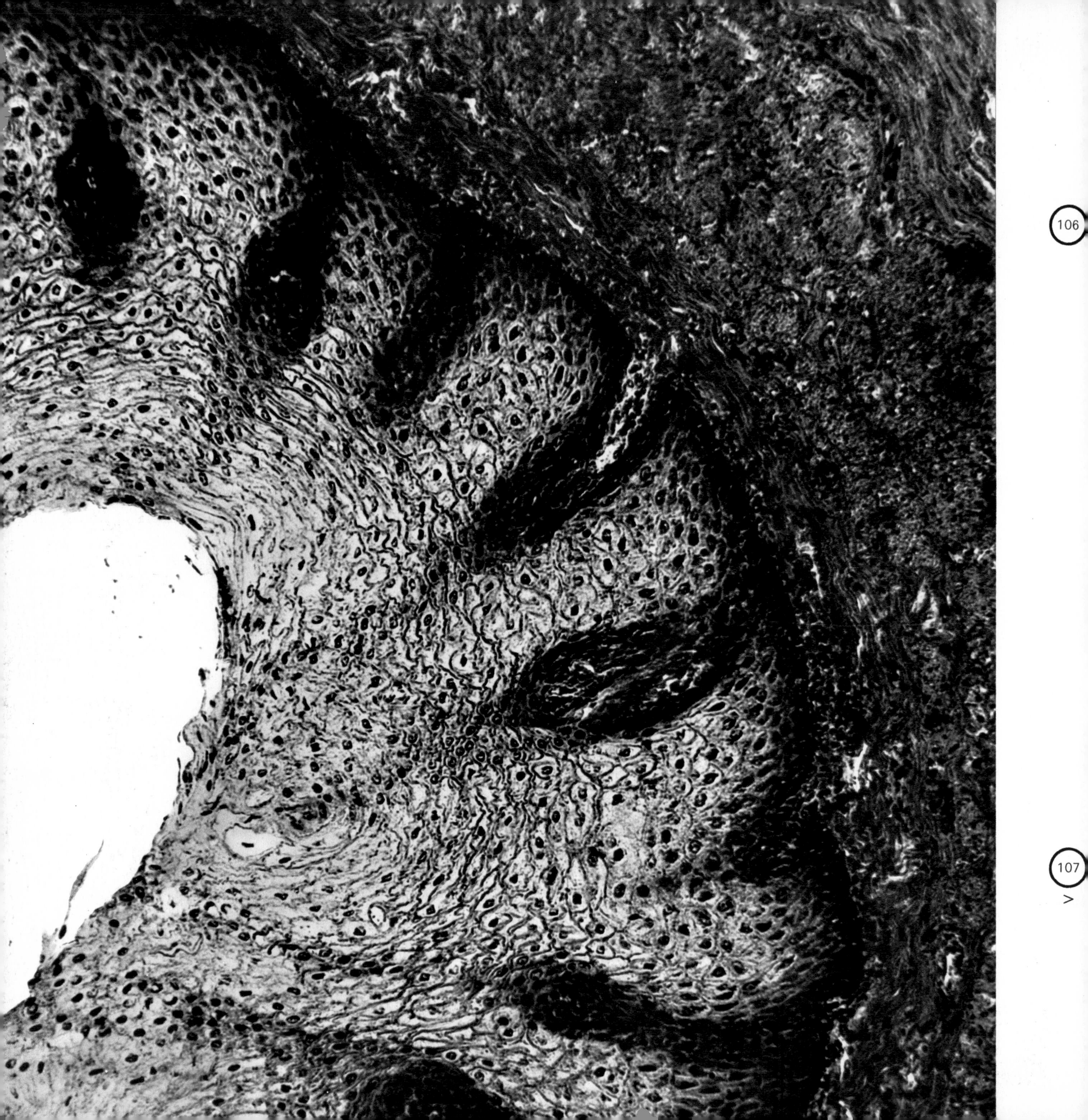
106
107

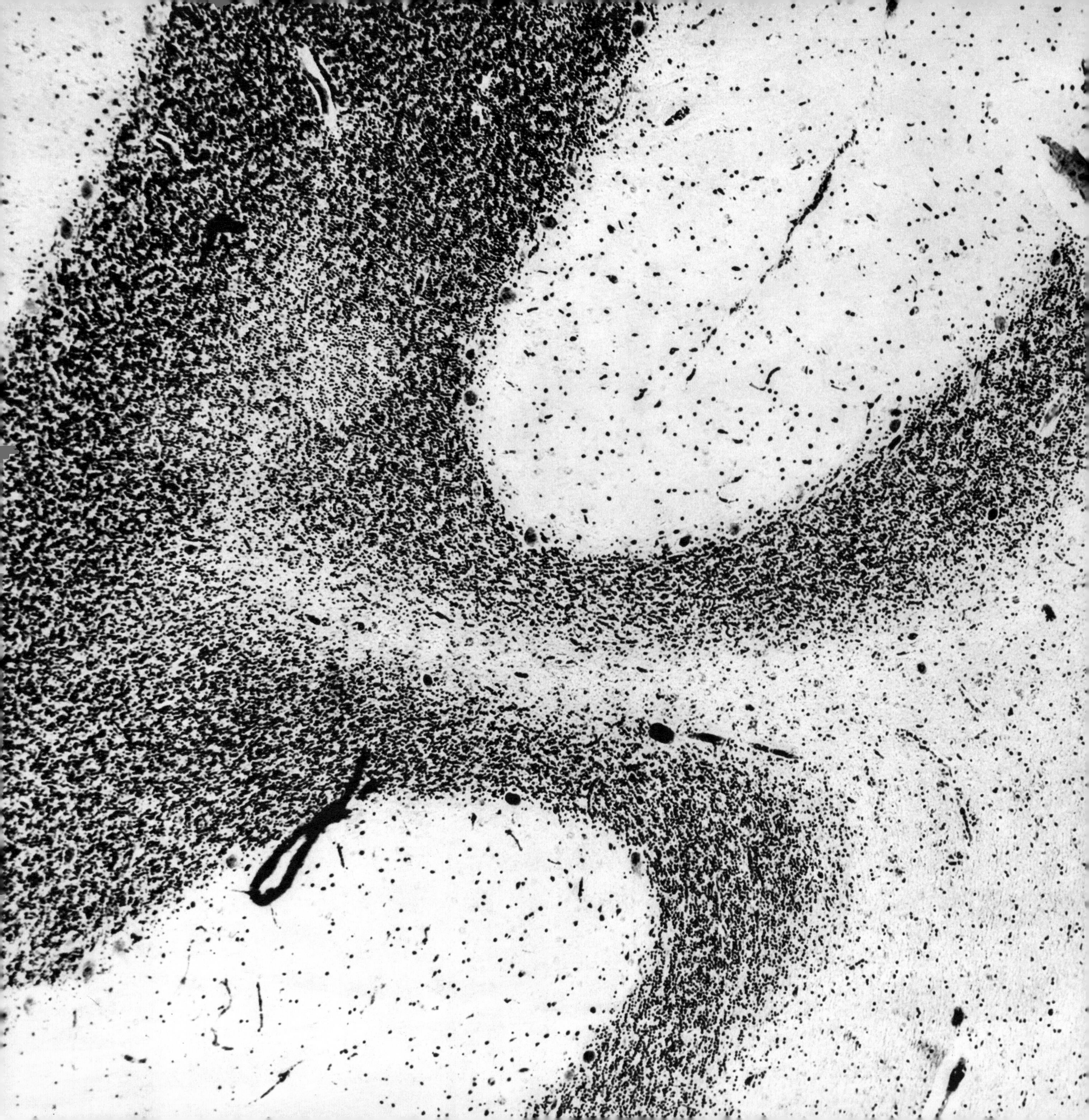

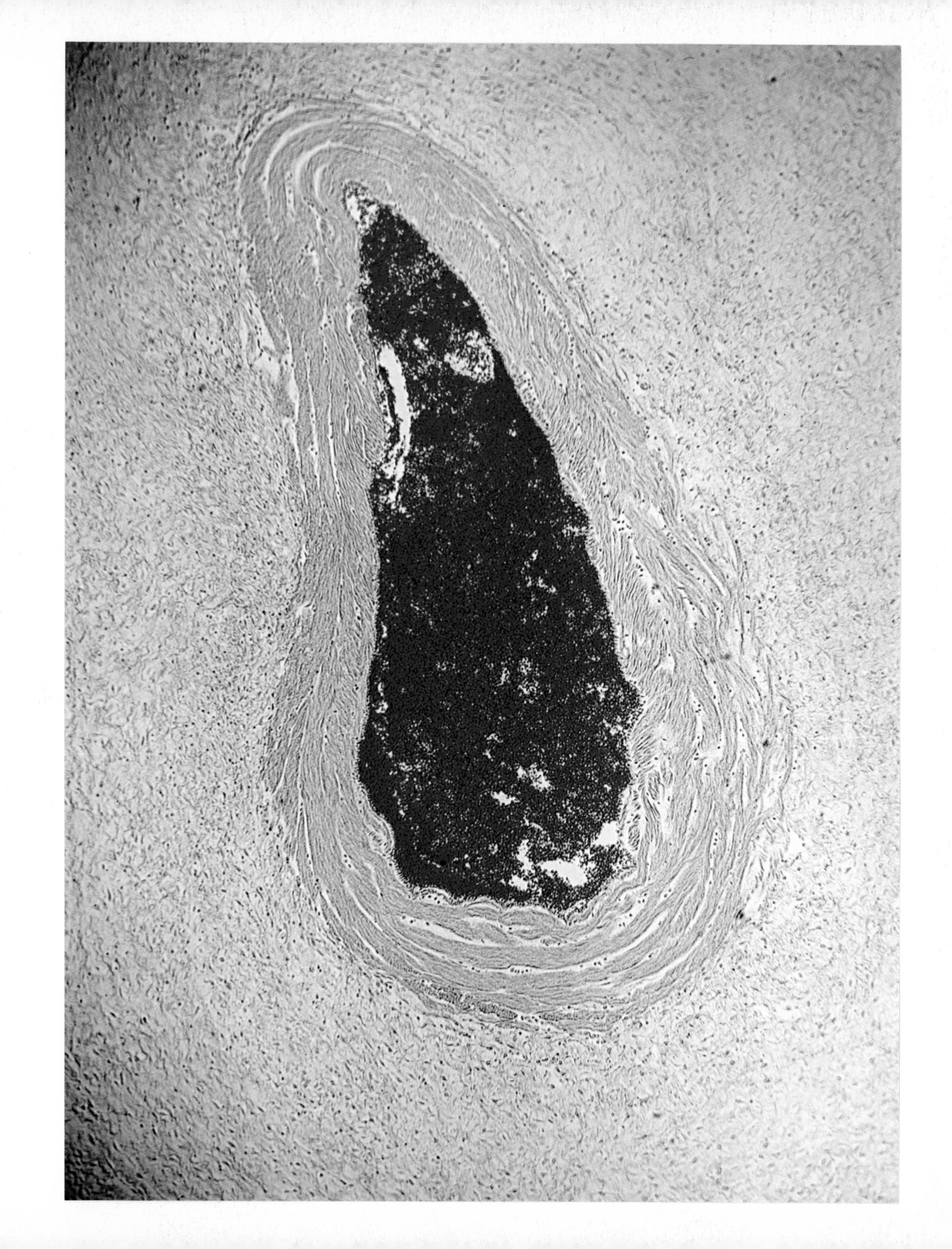

108

109

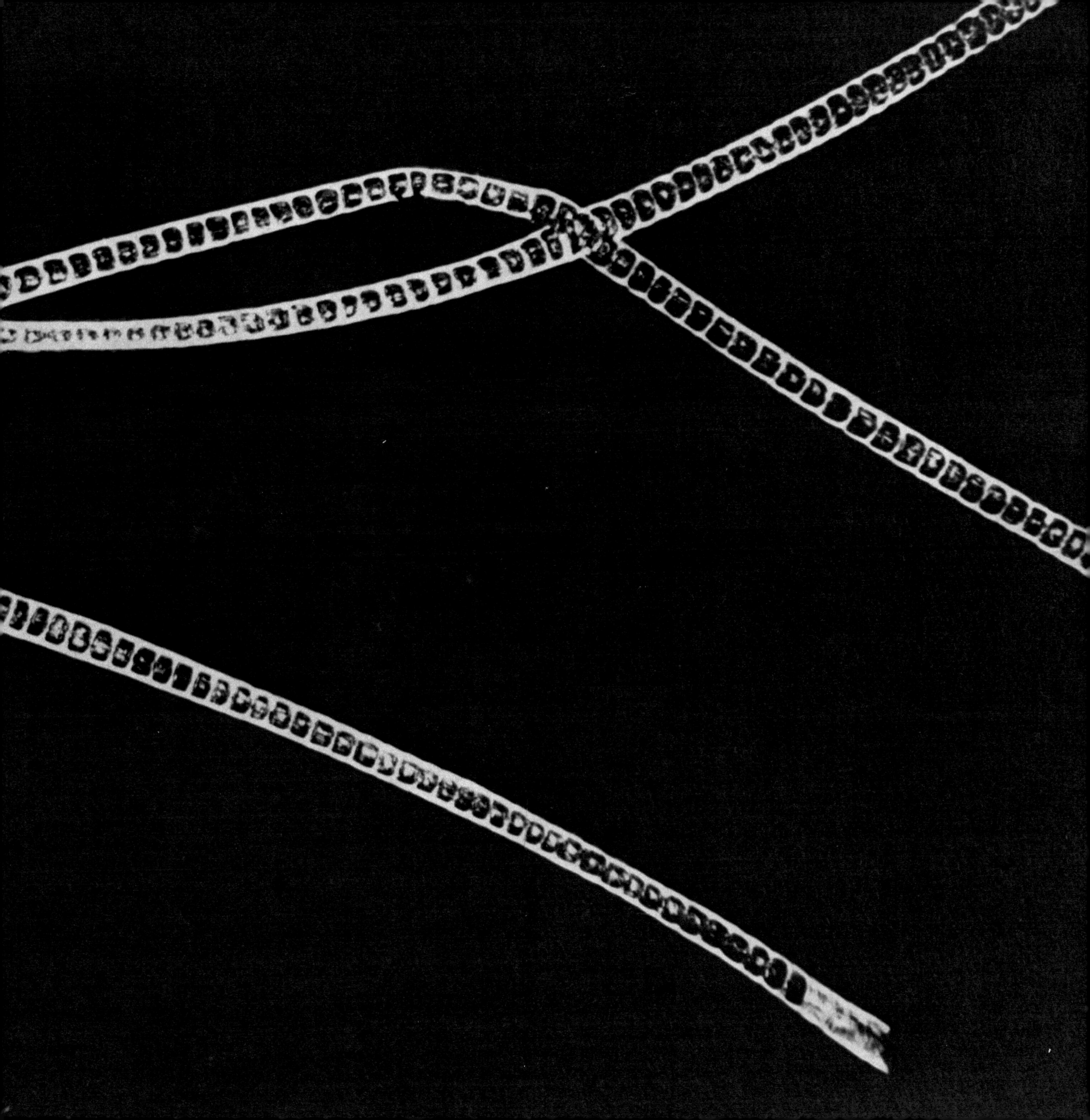

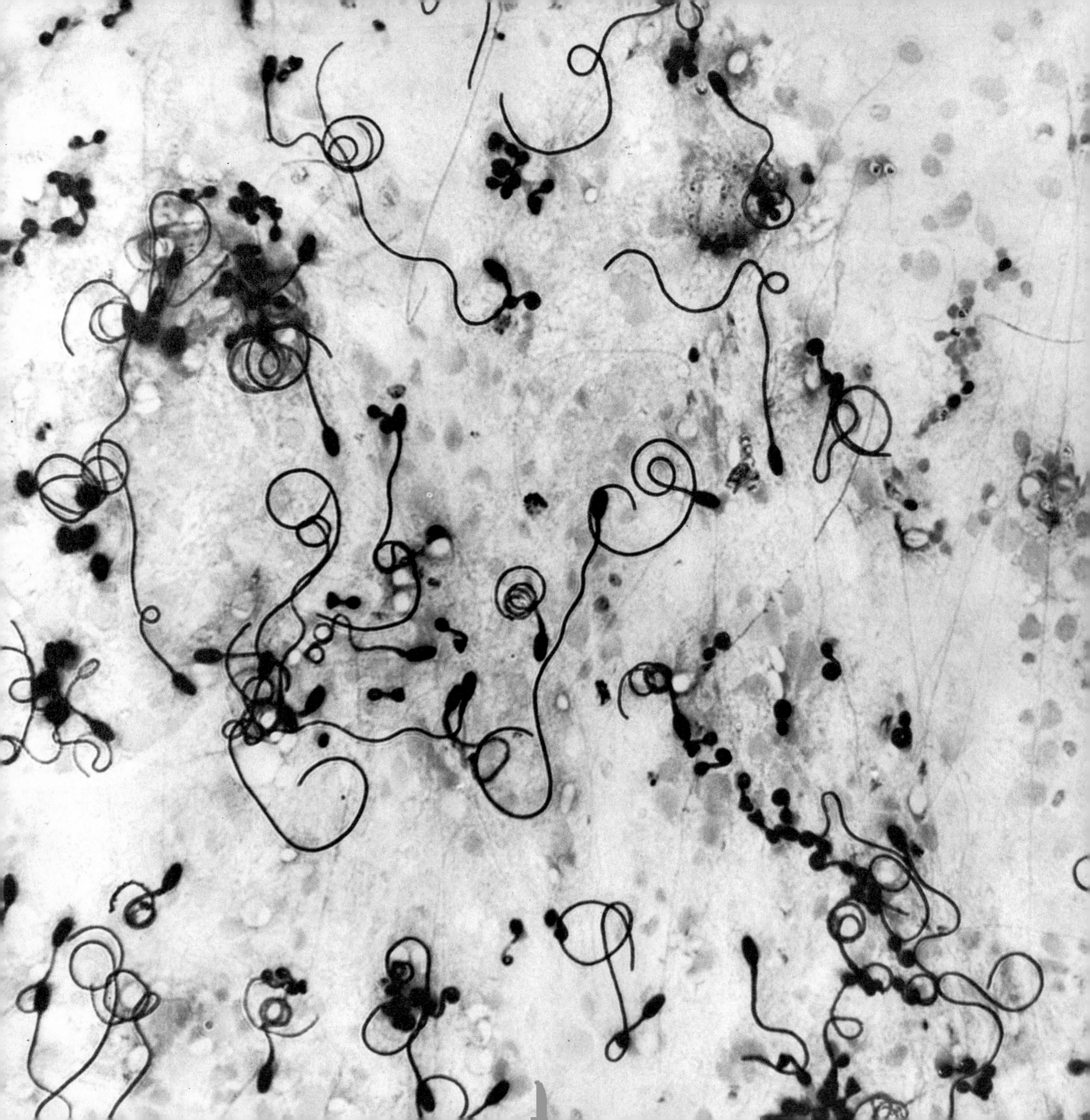

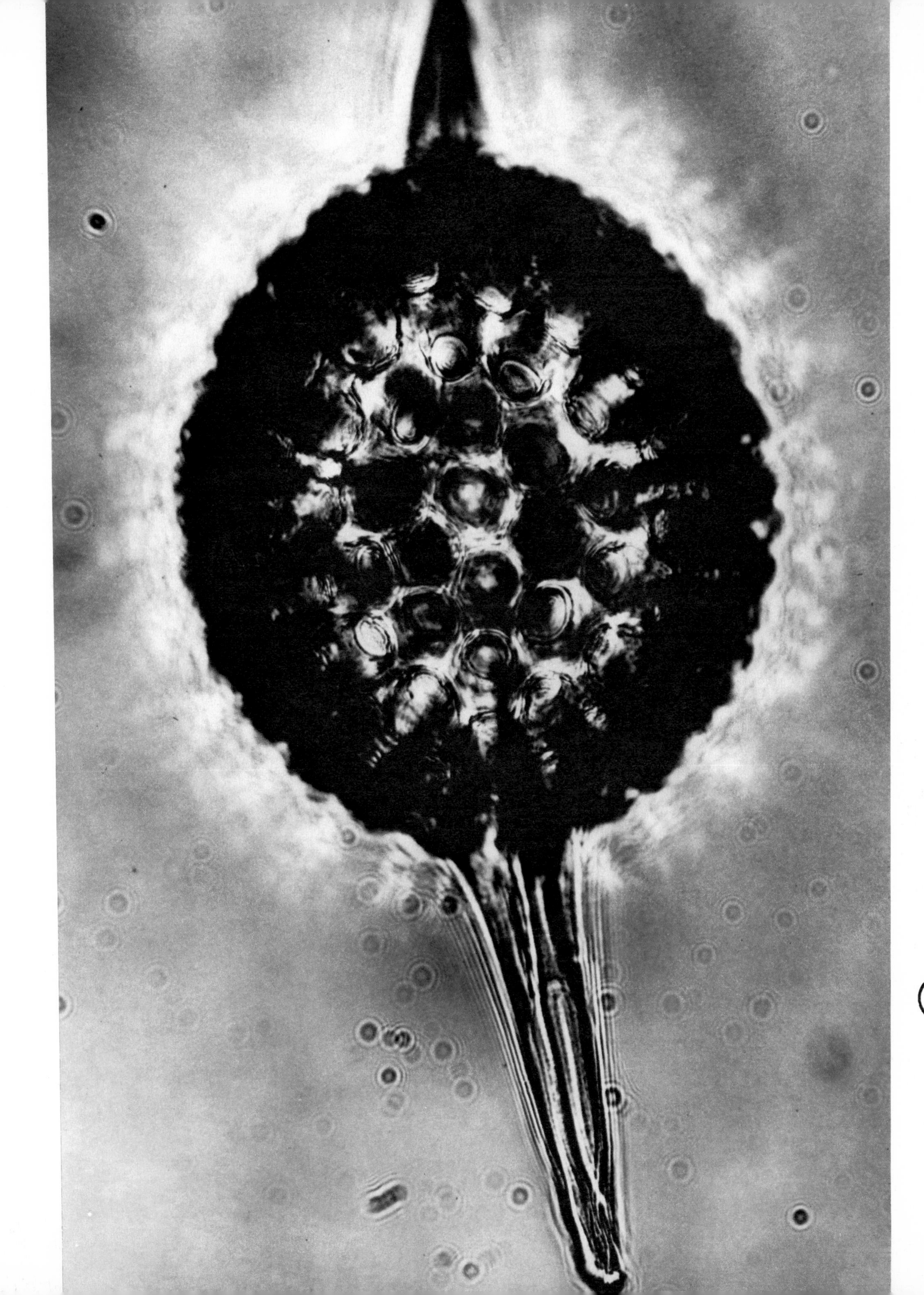

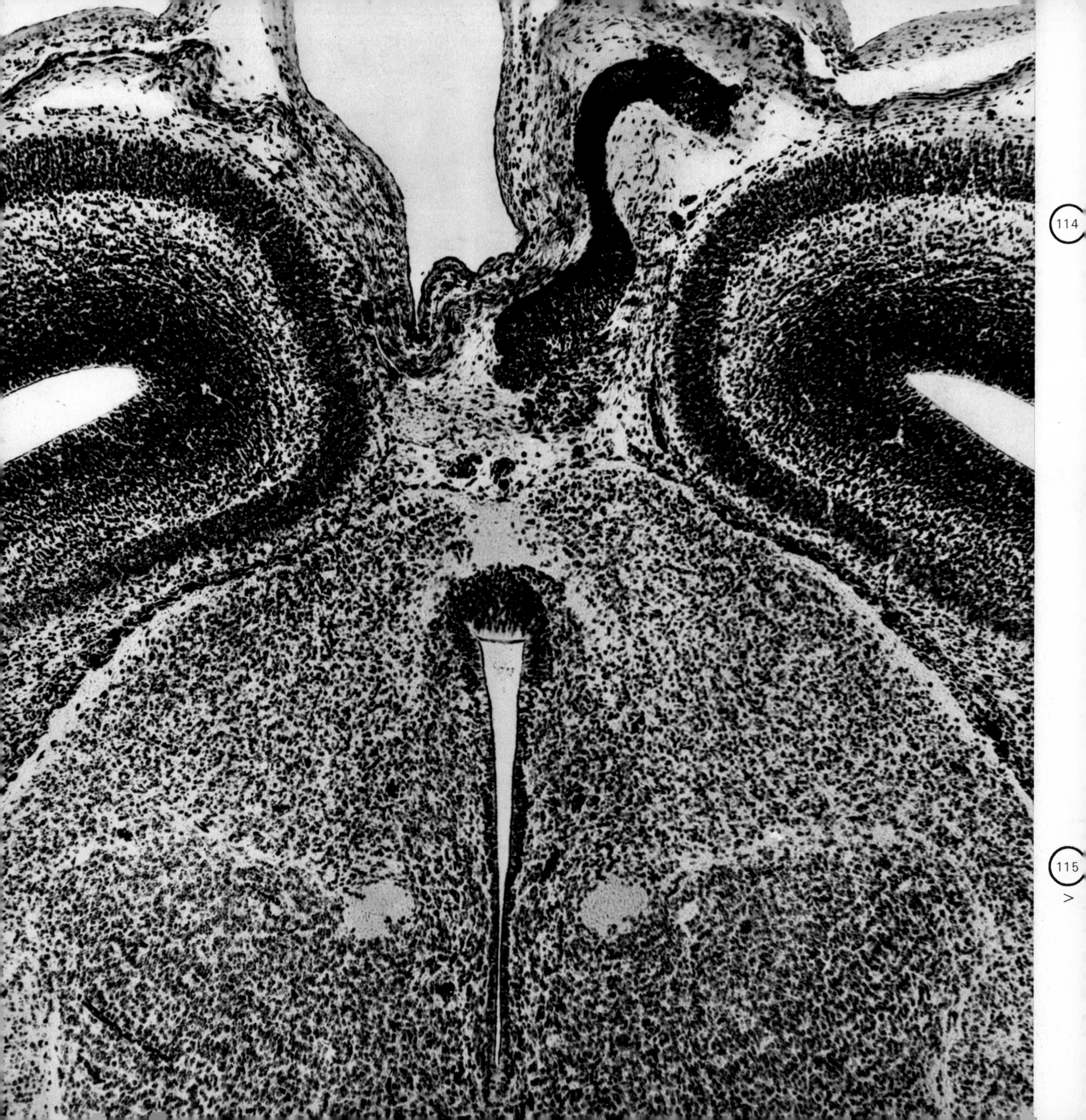
114
115

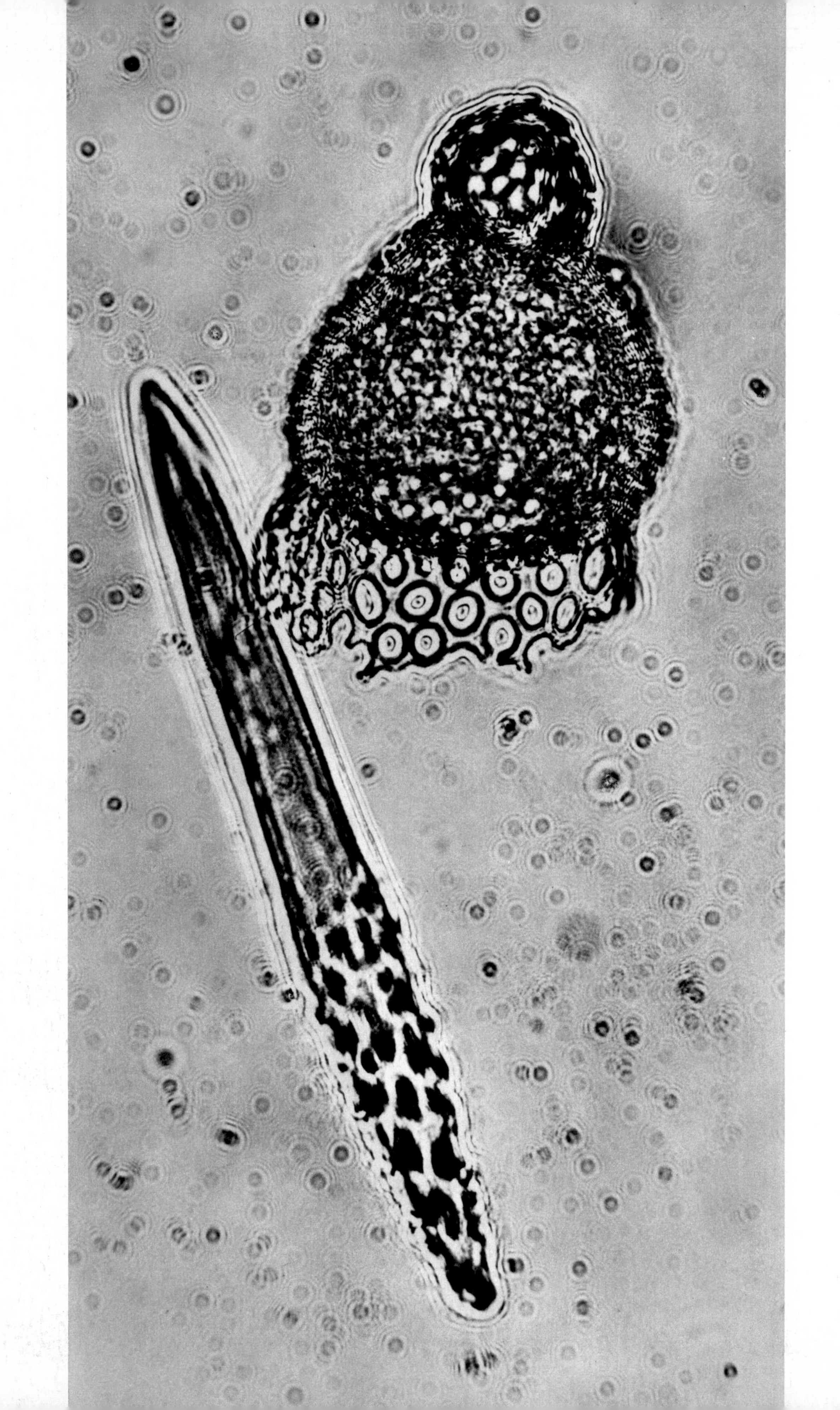

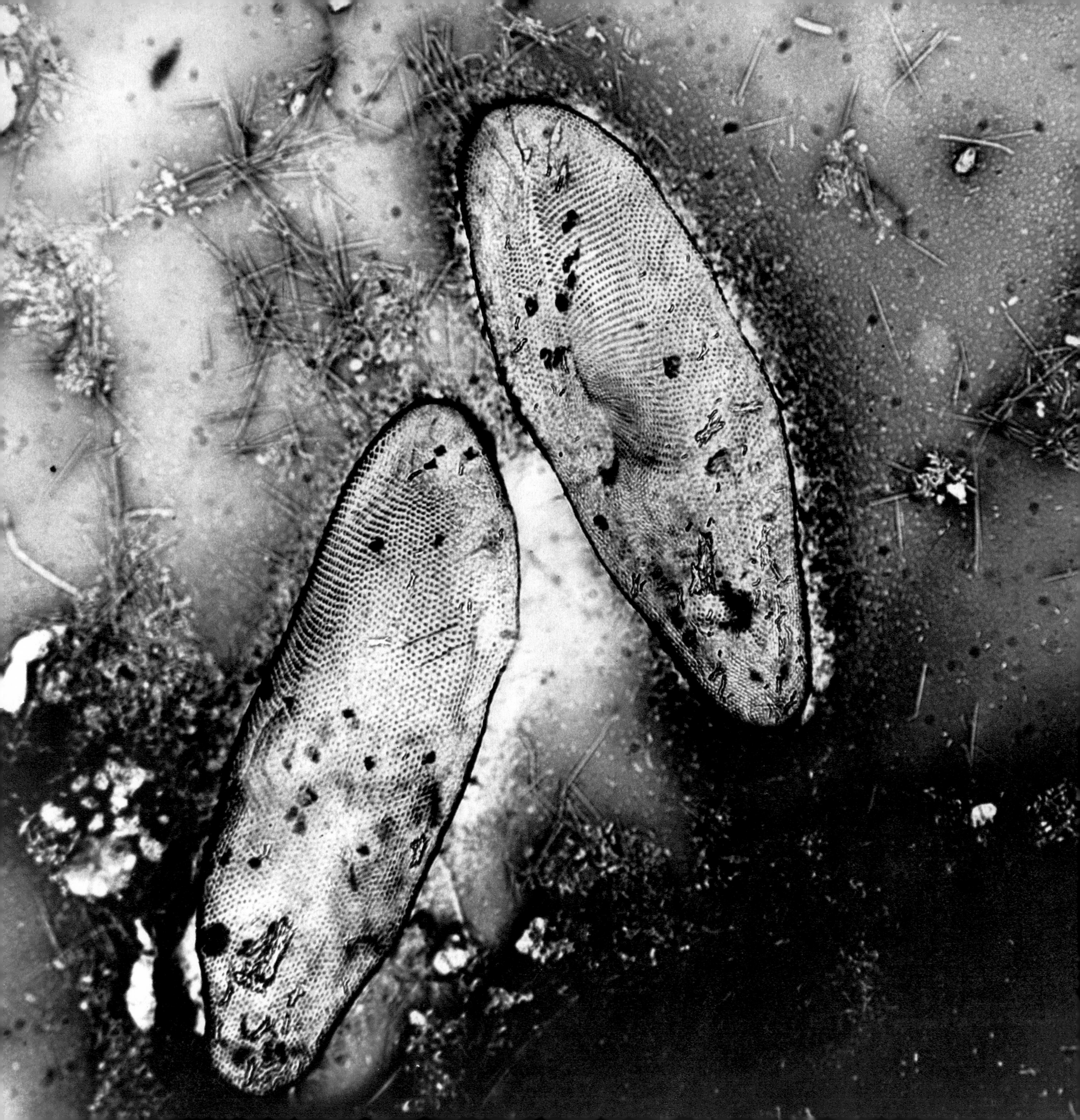

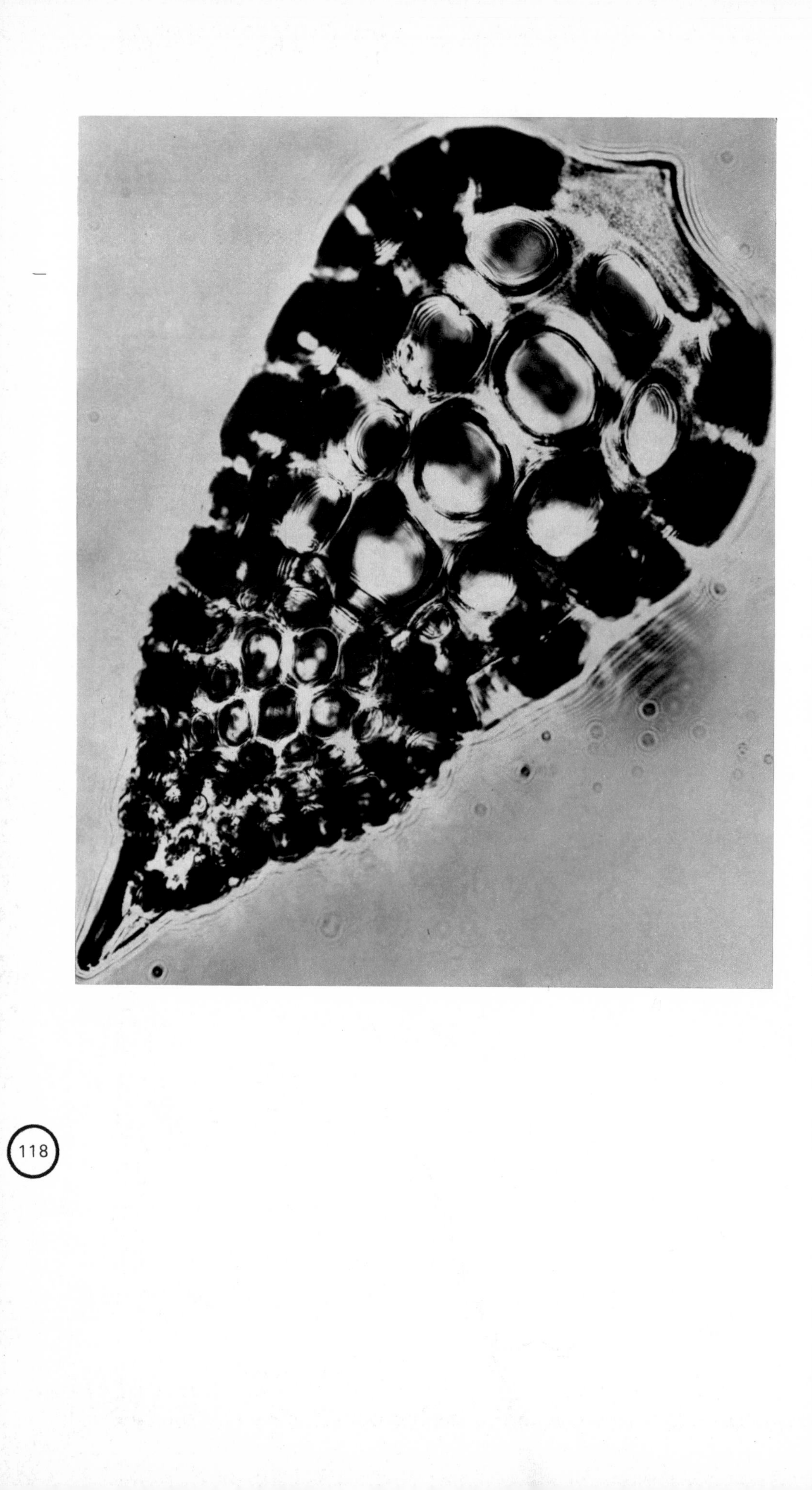

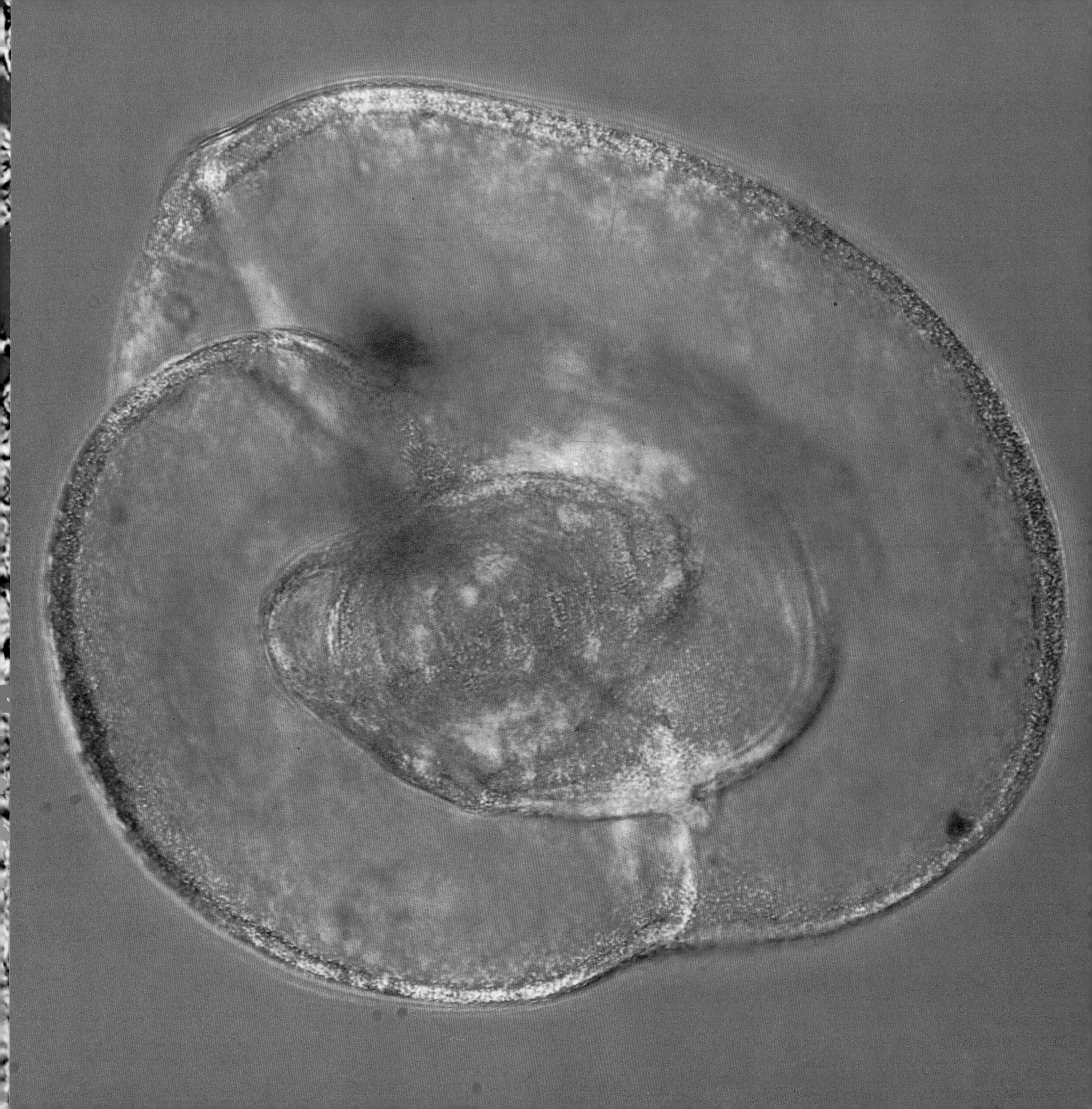

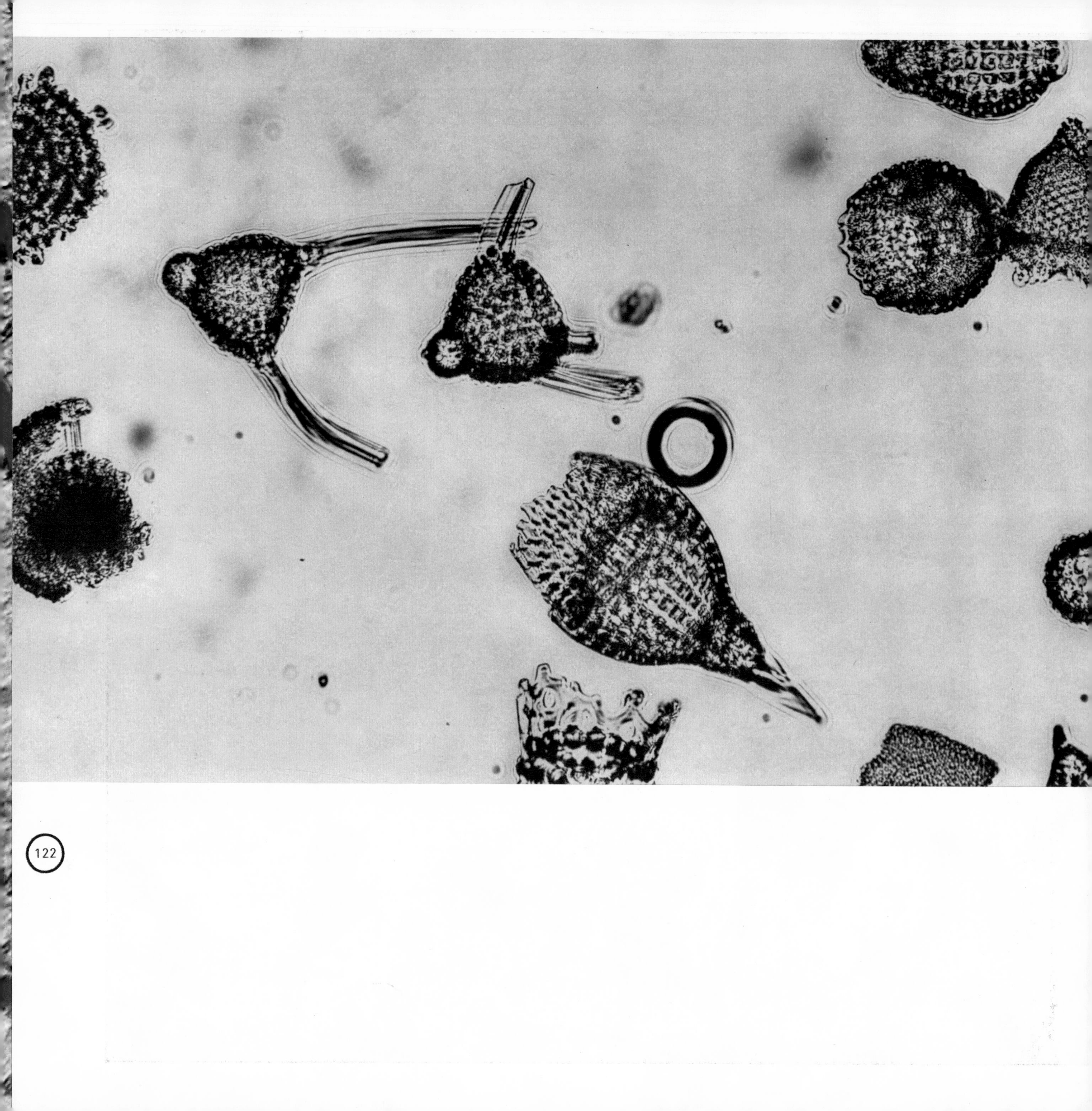

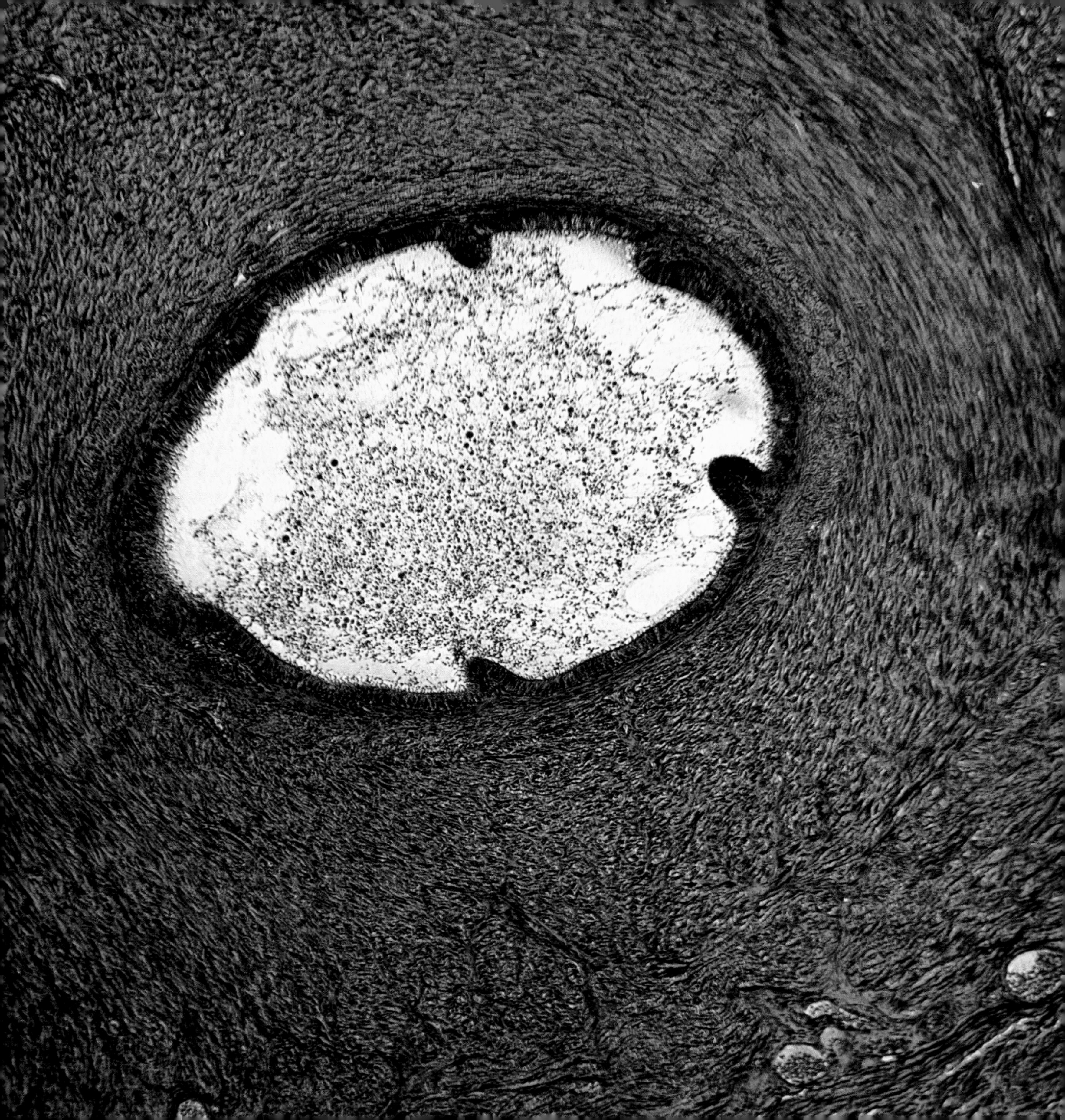

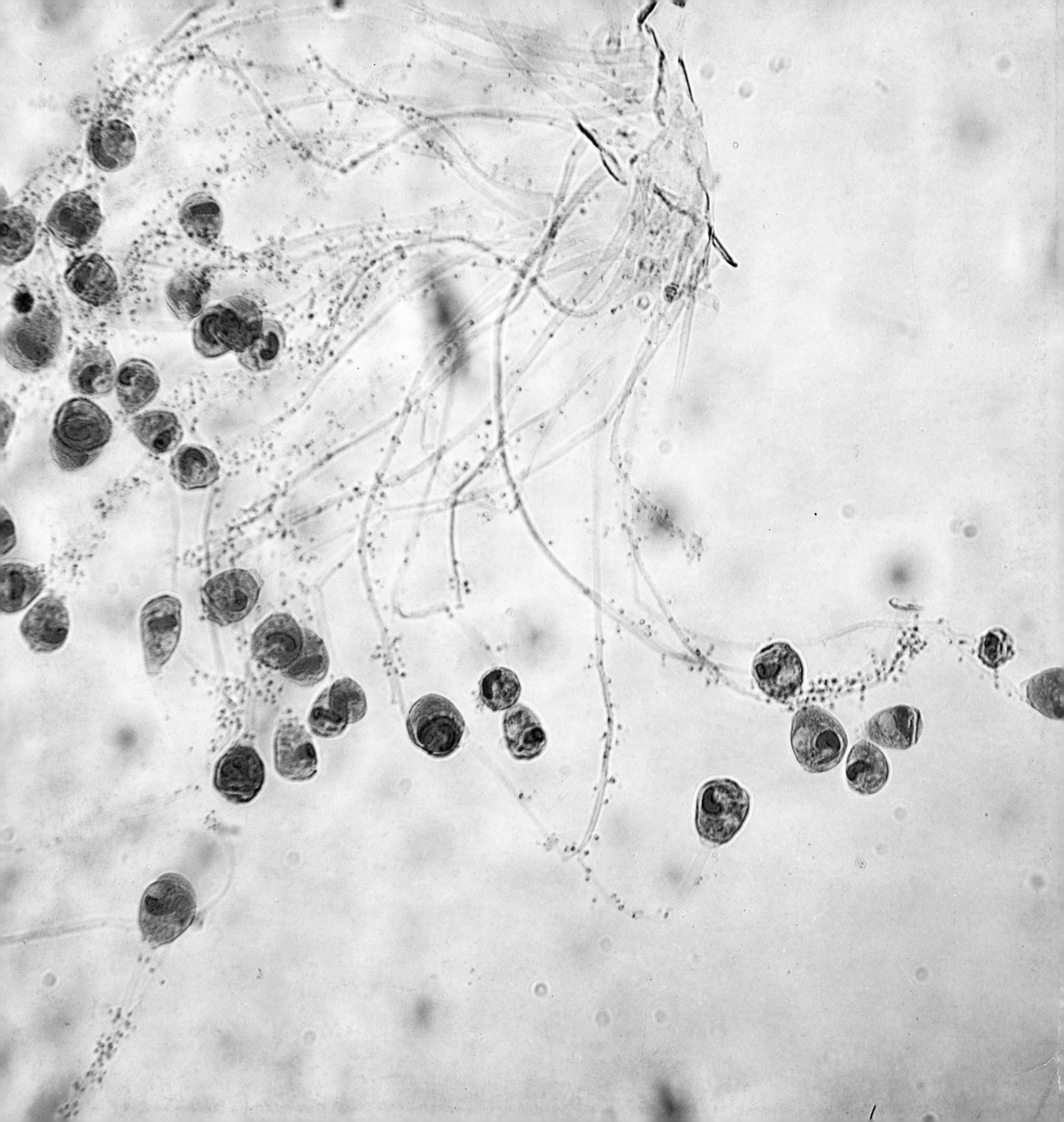

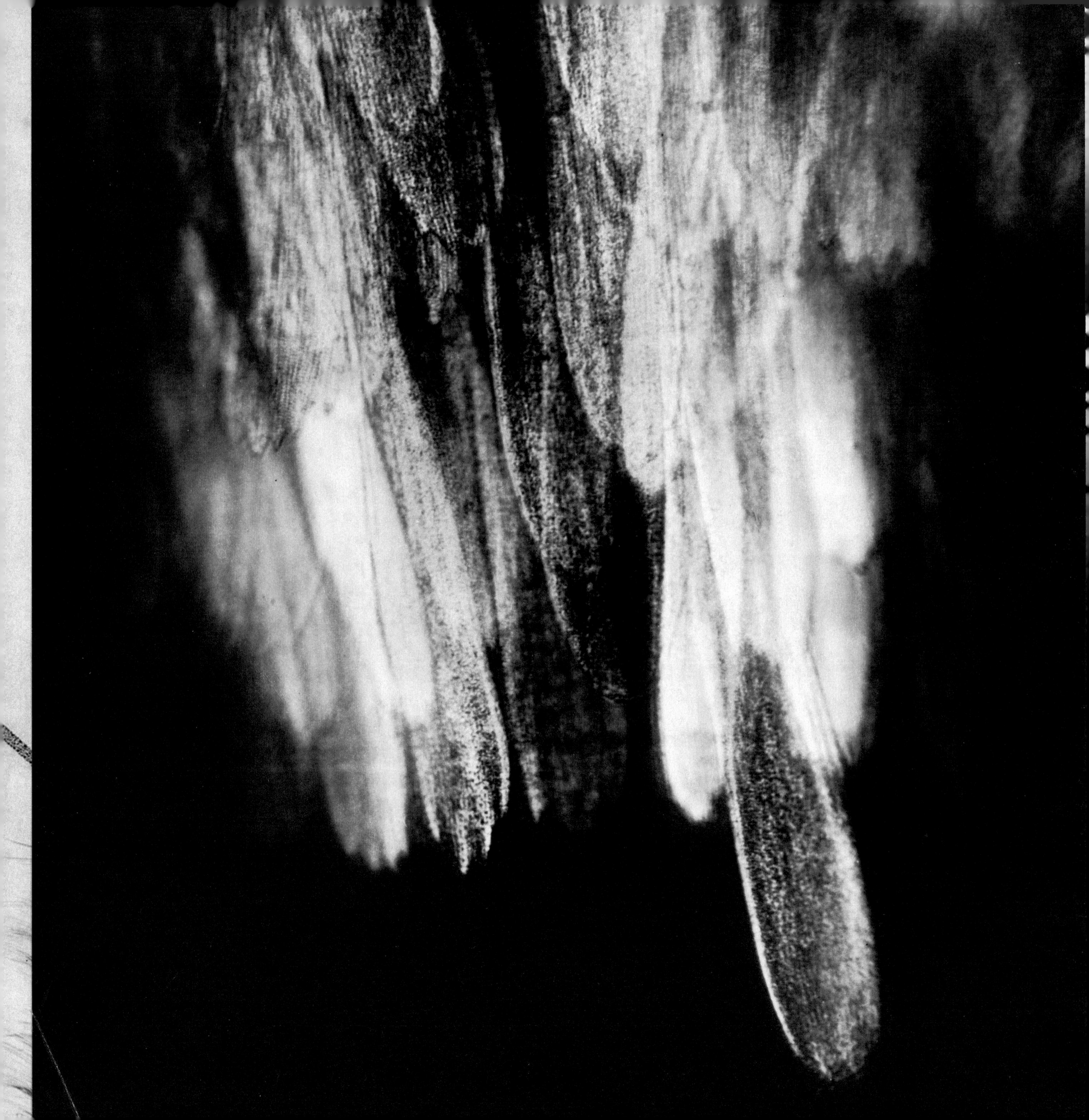

127
128
129

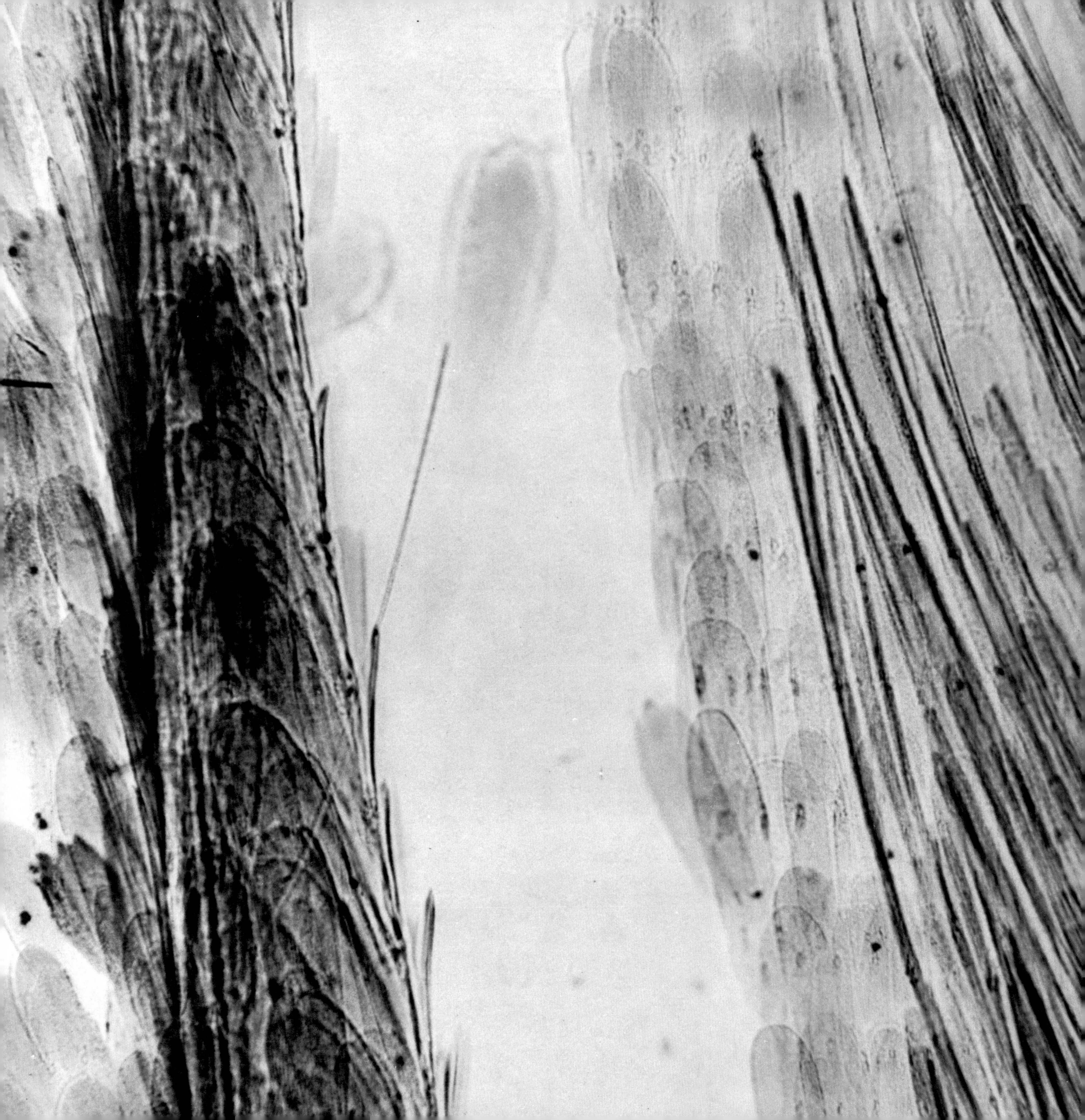

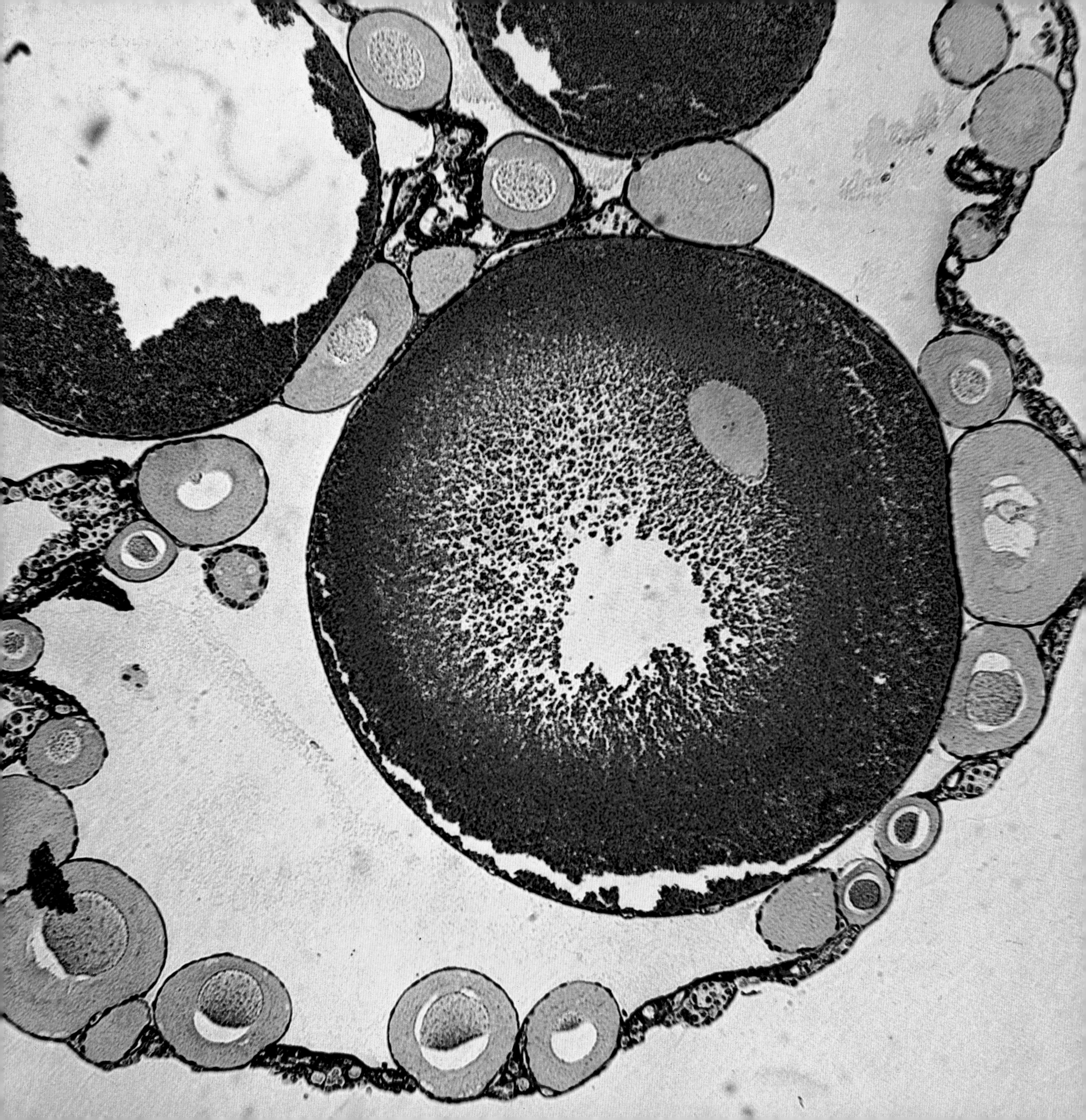

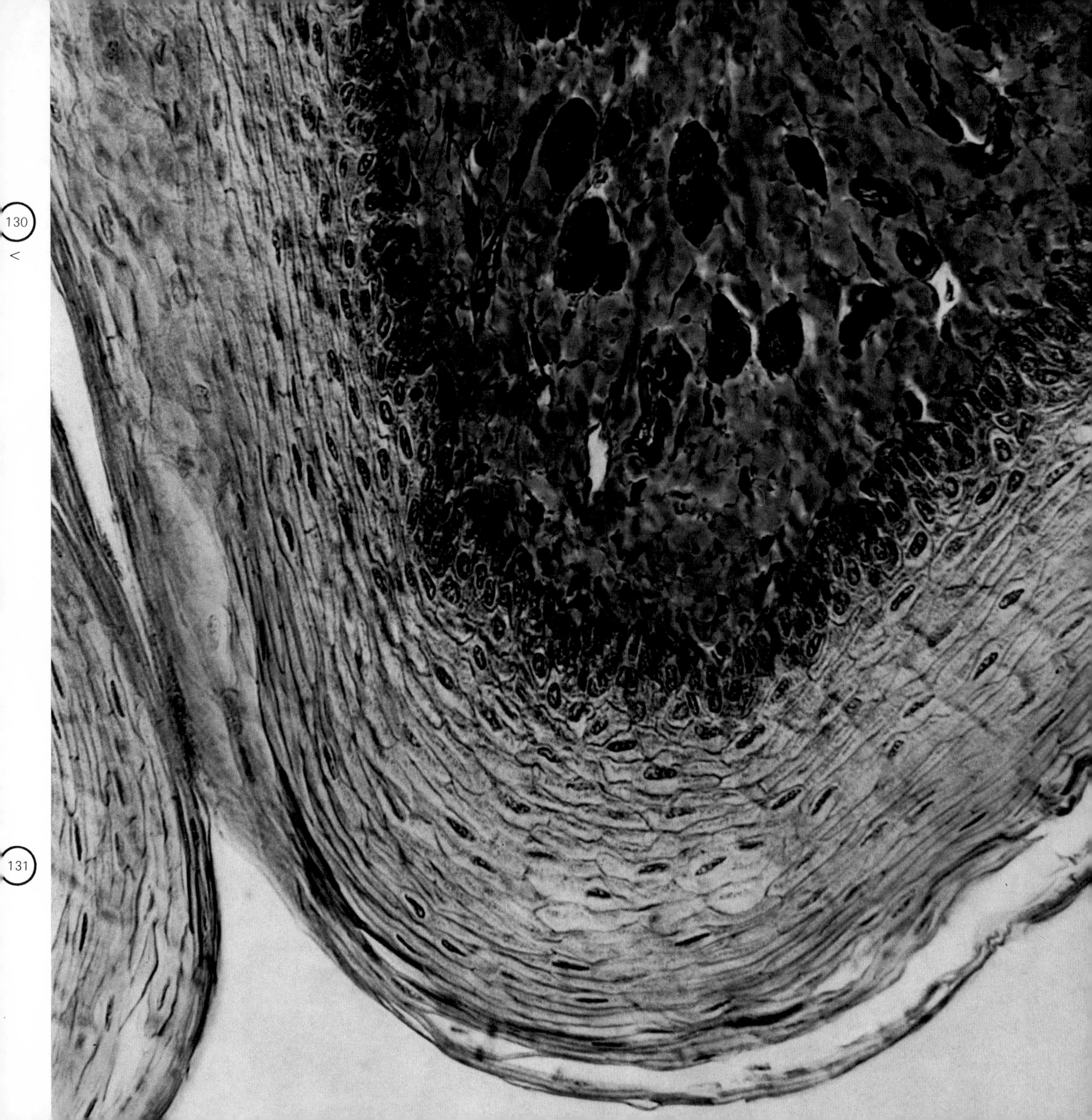
130
131

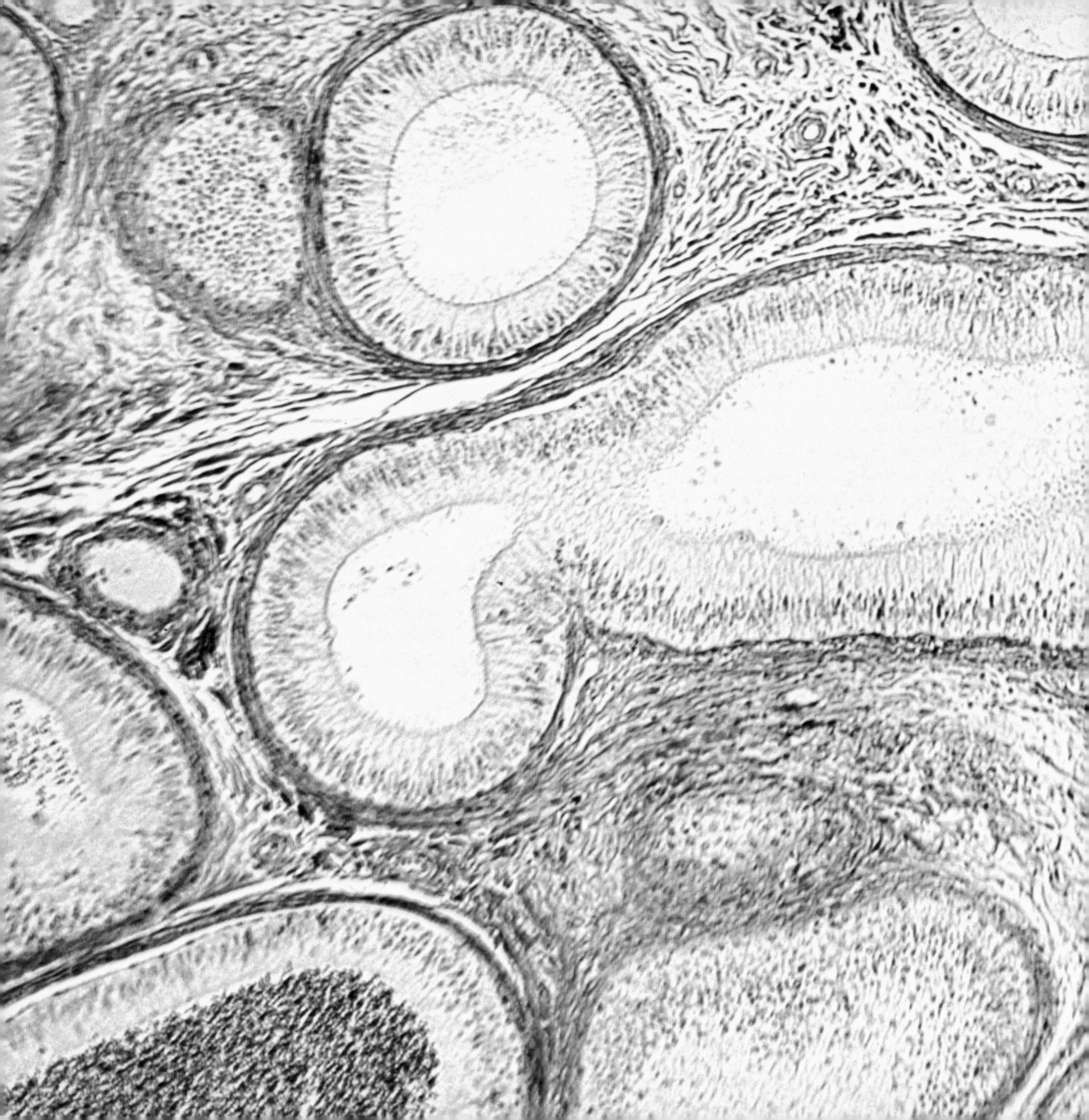

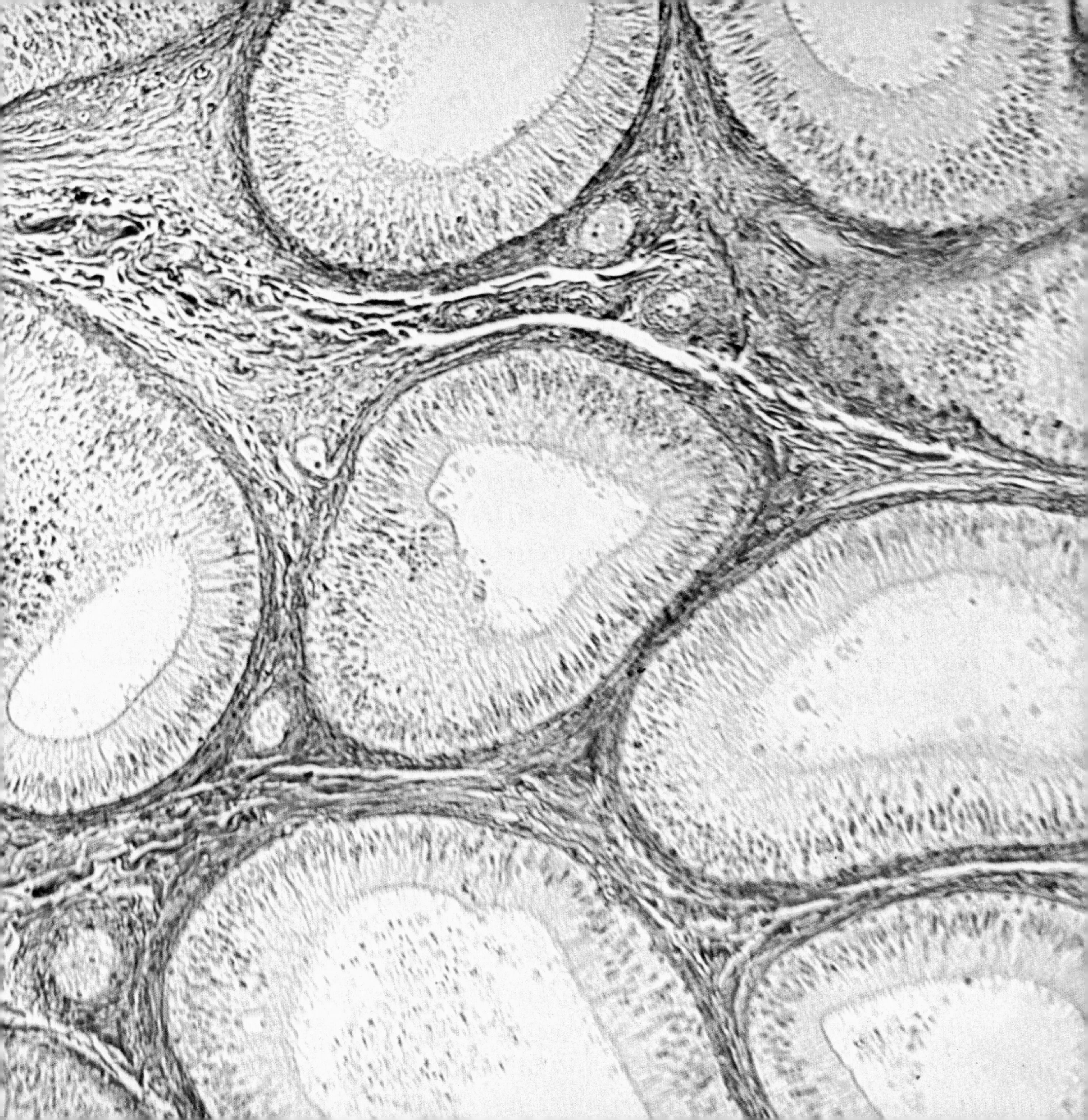

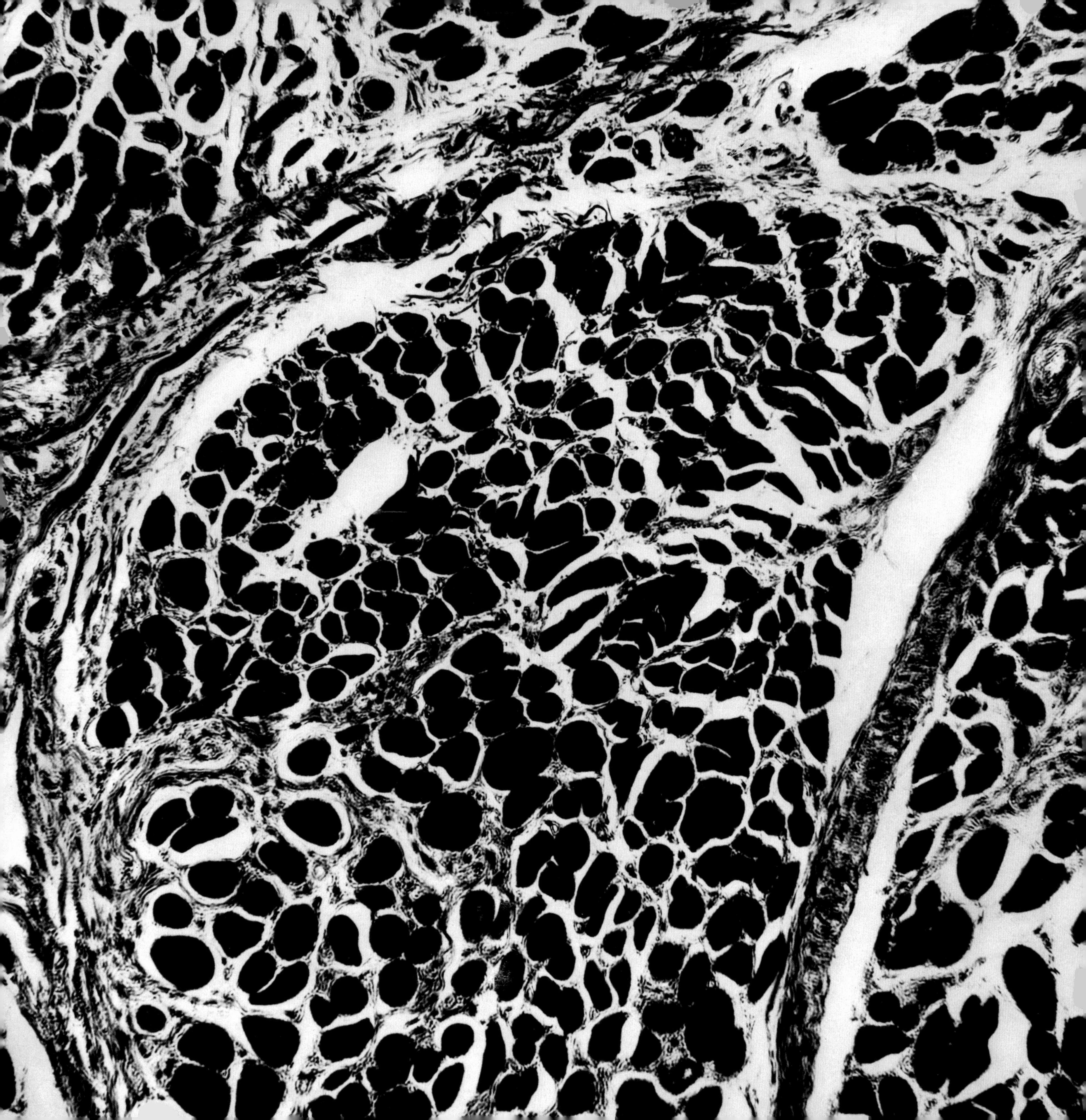

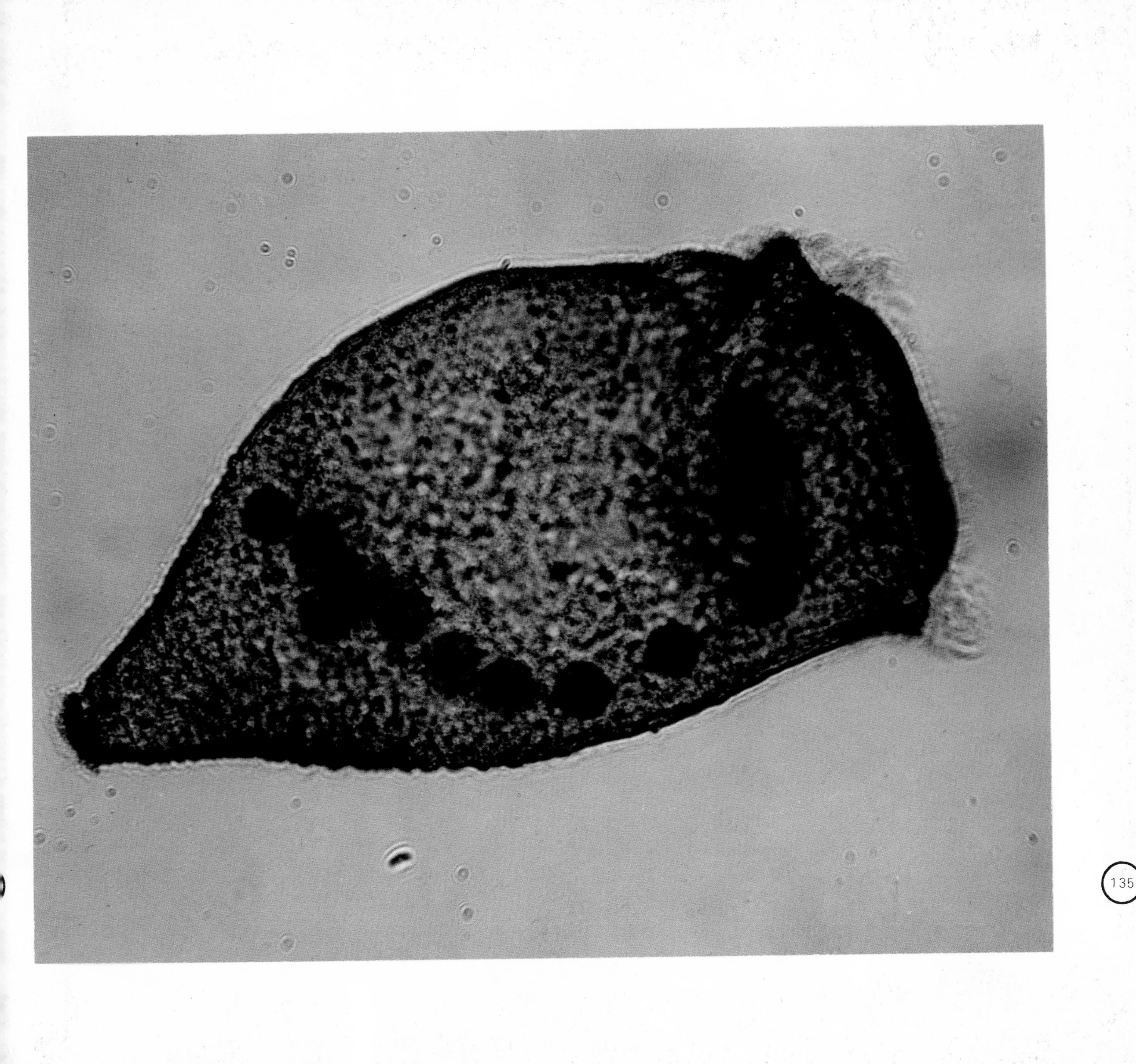
135

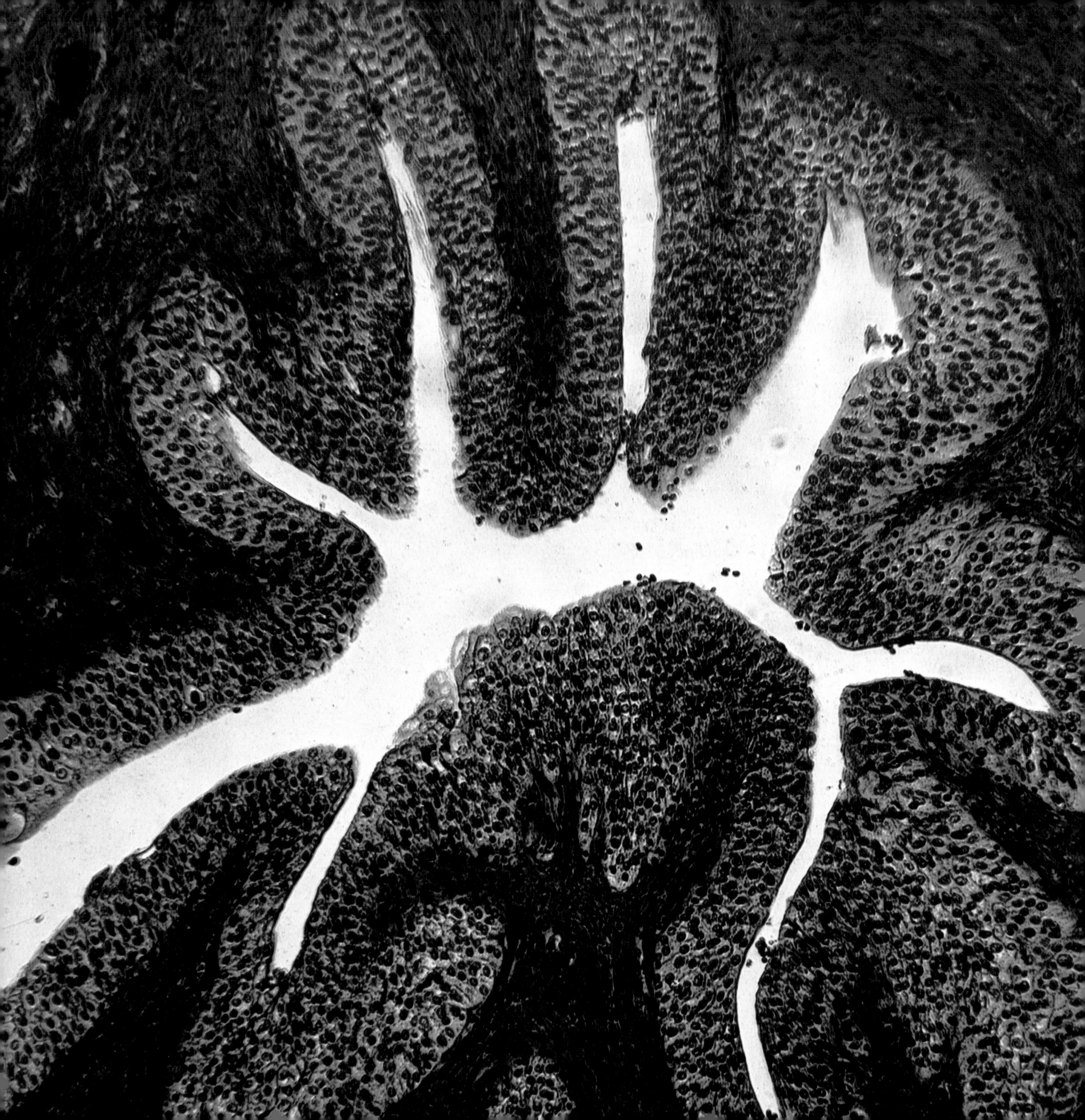

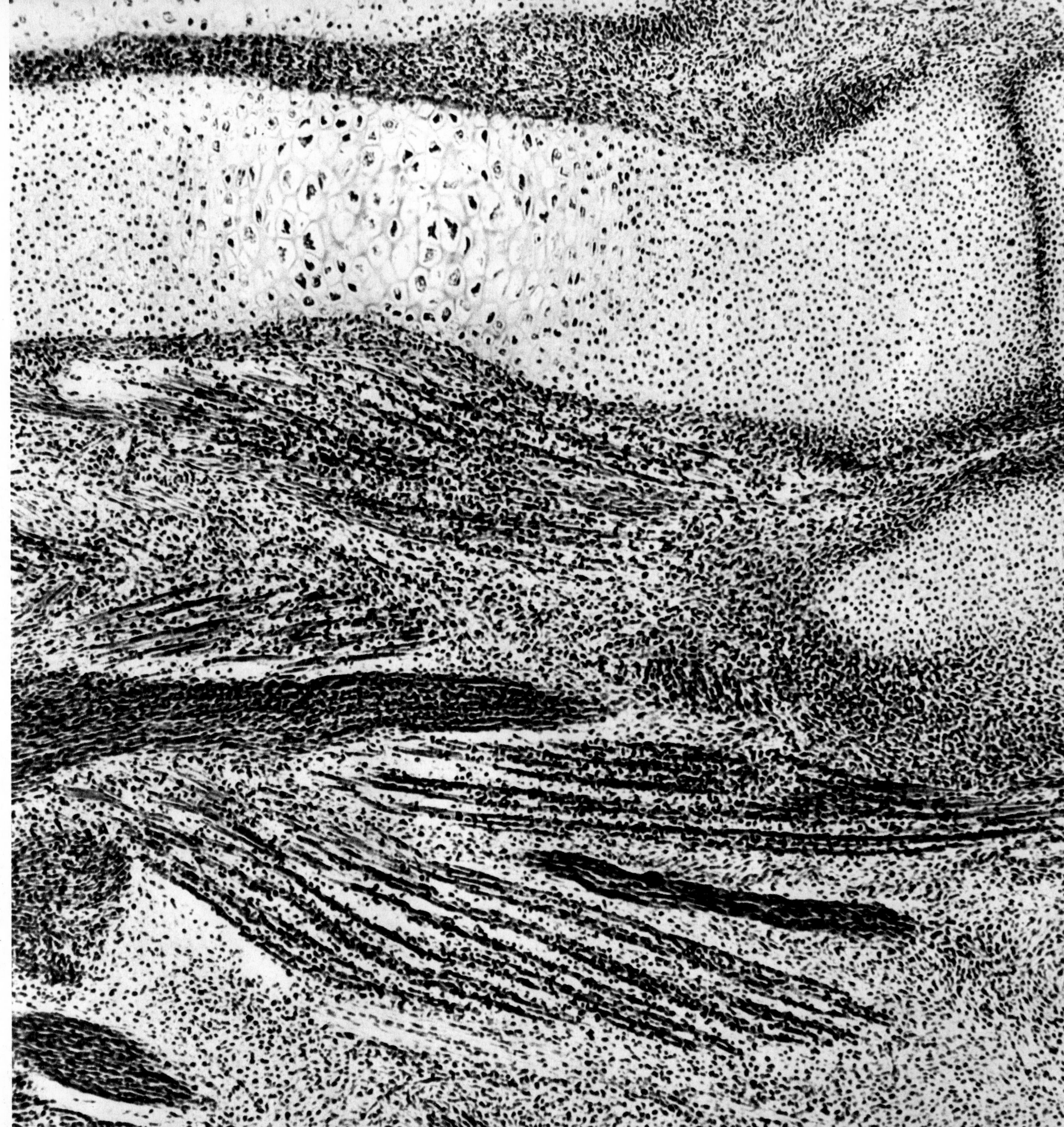

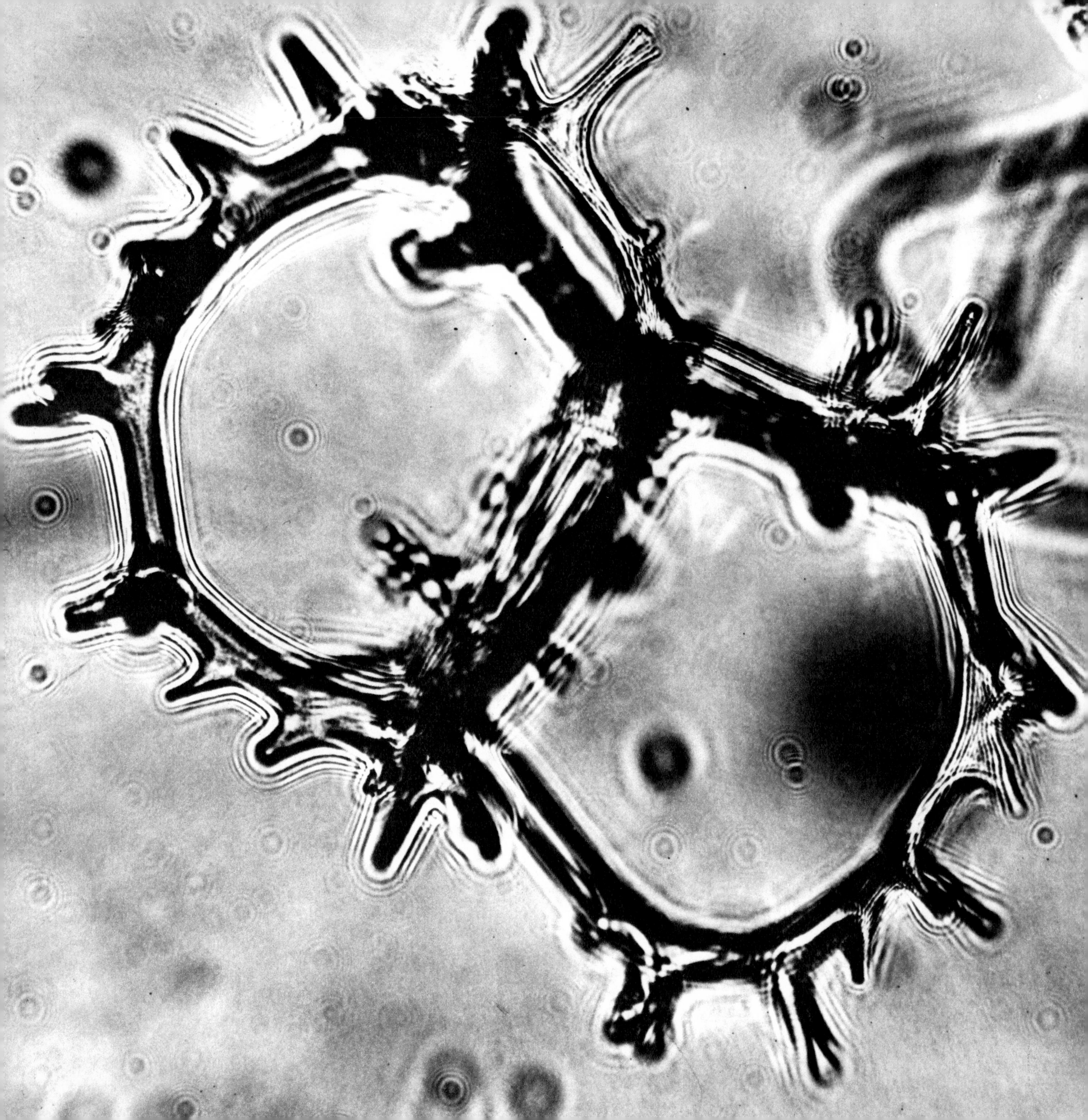

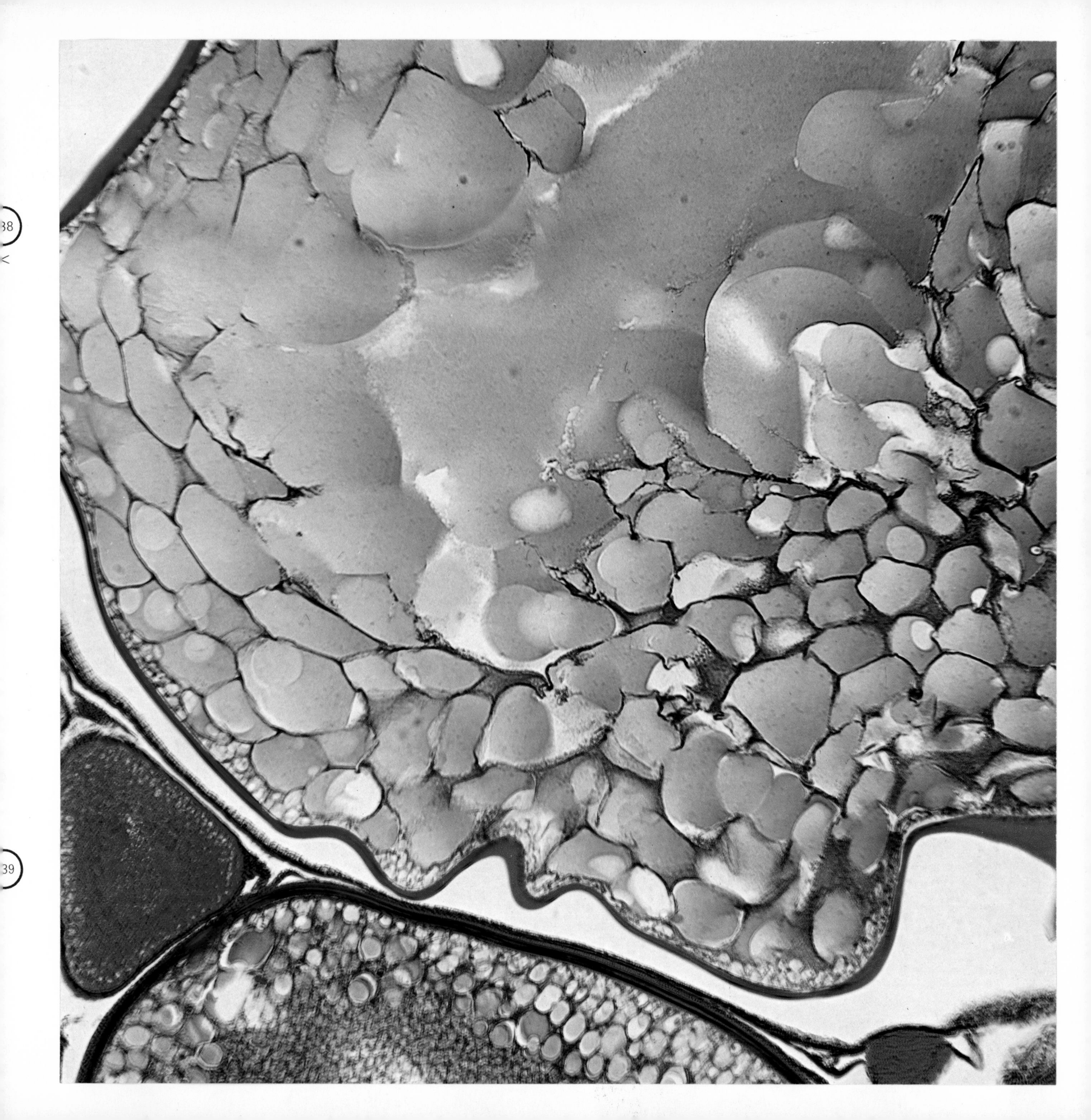

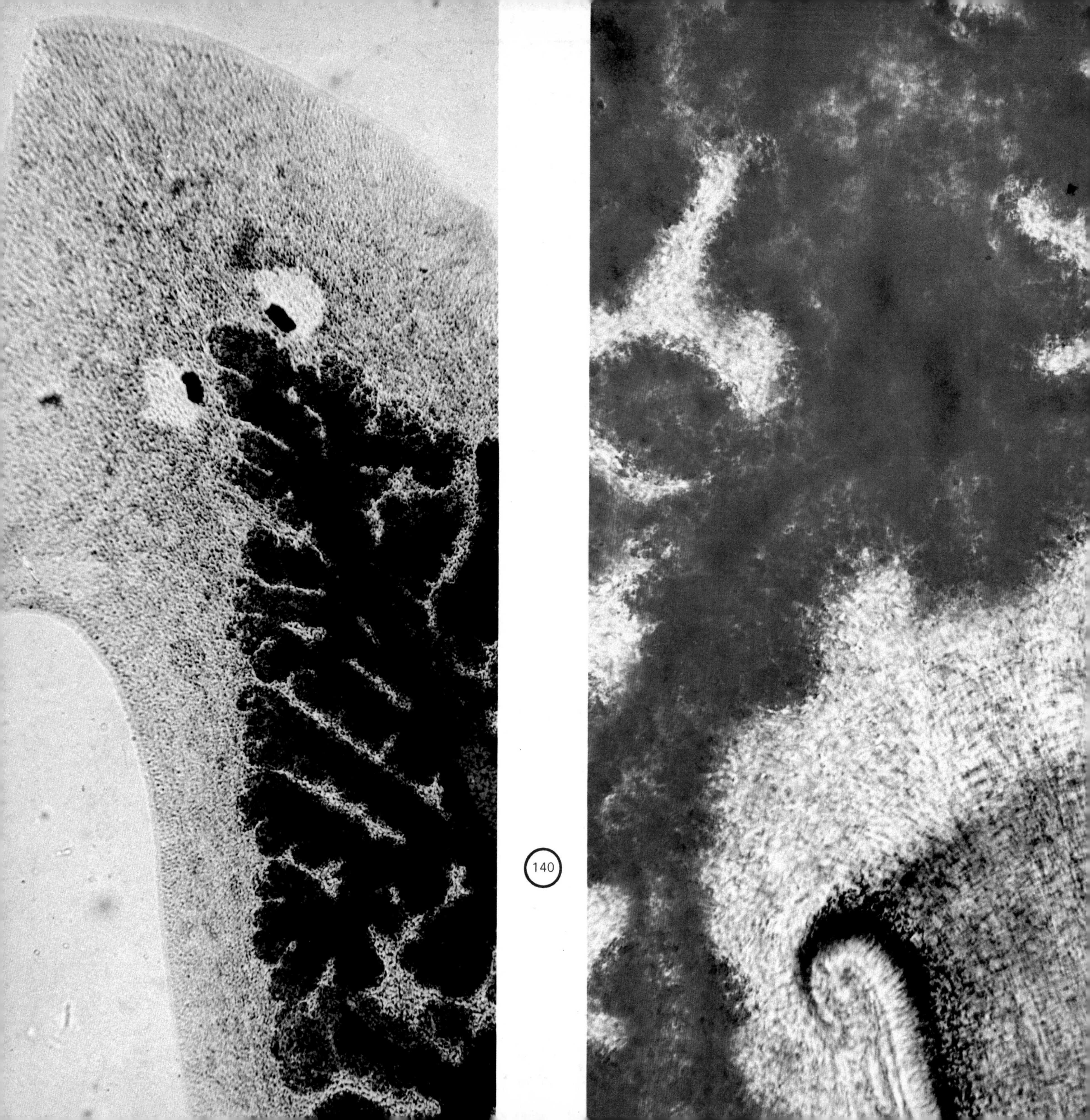

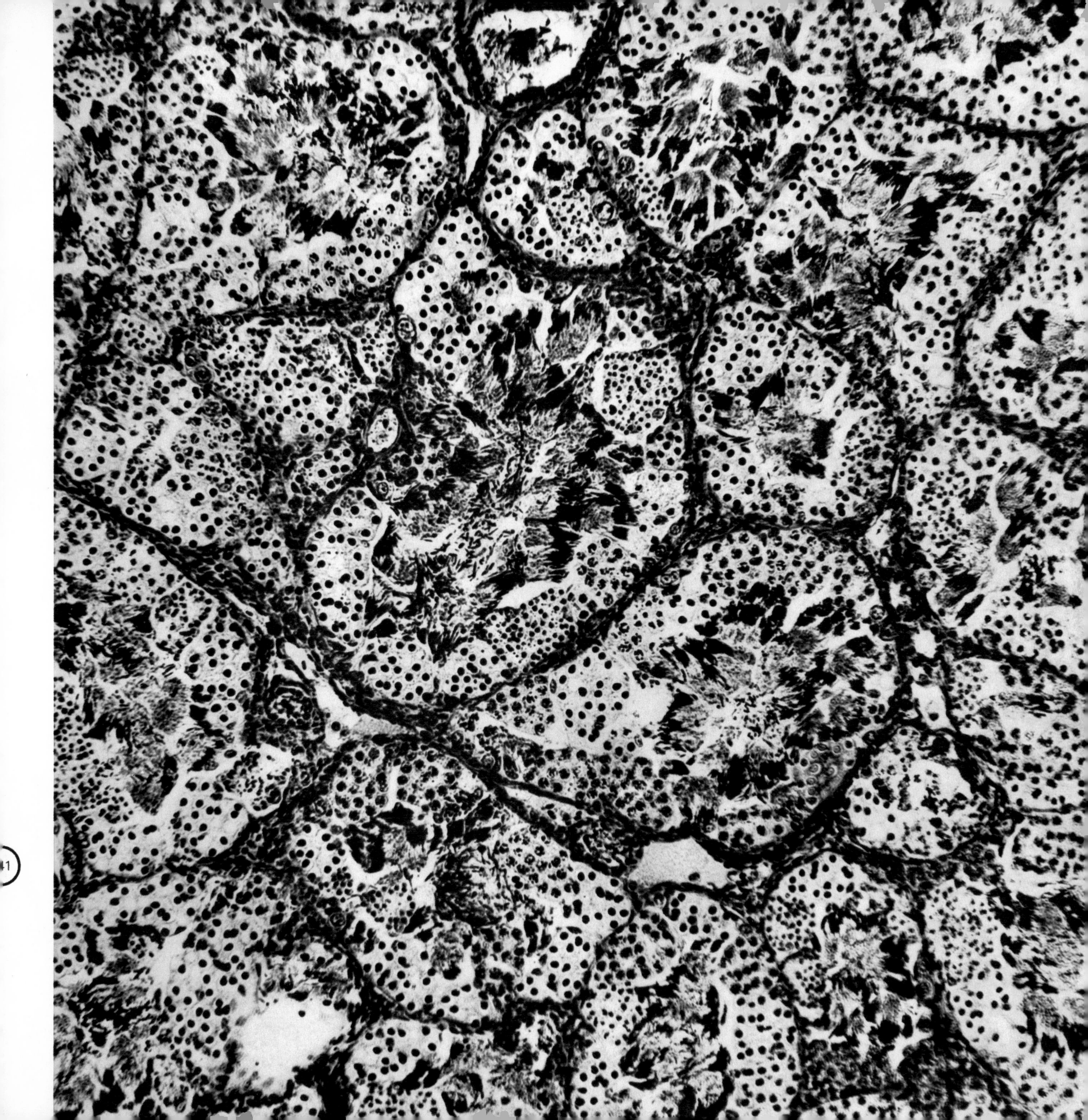

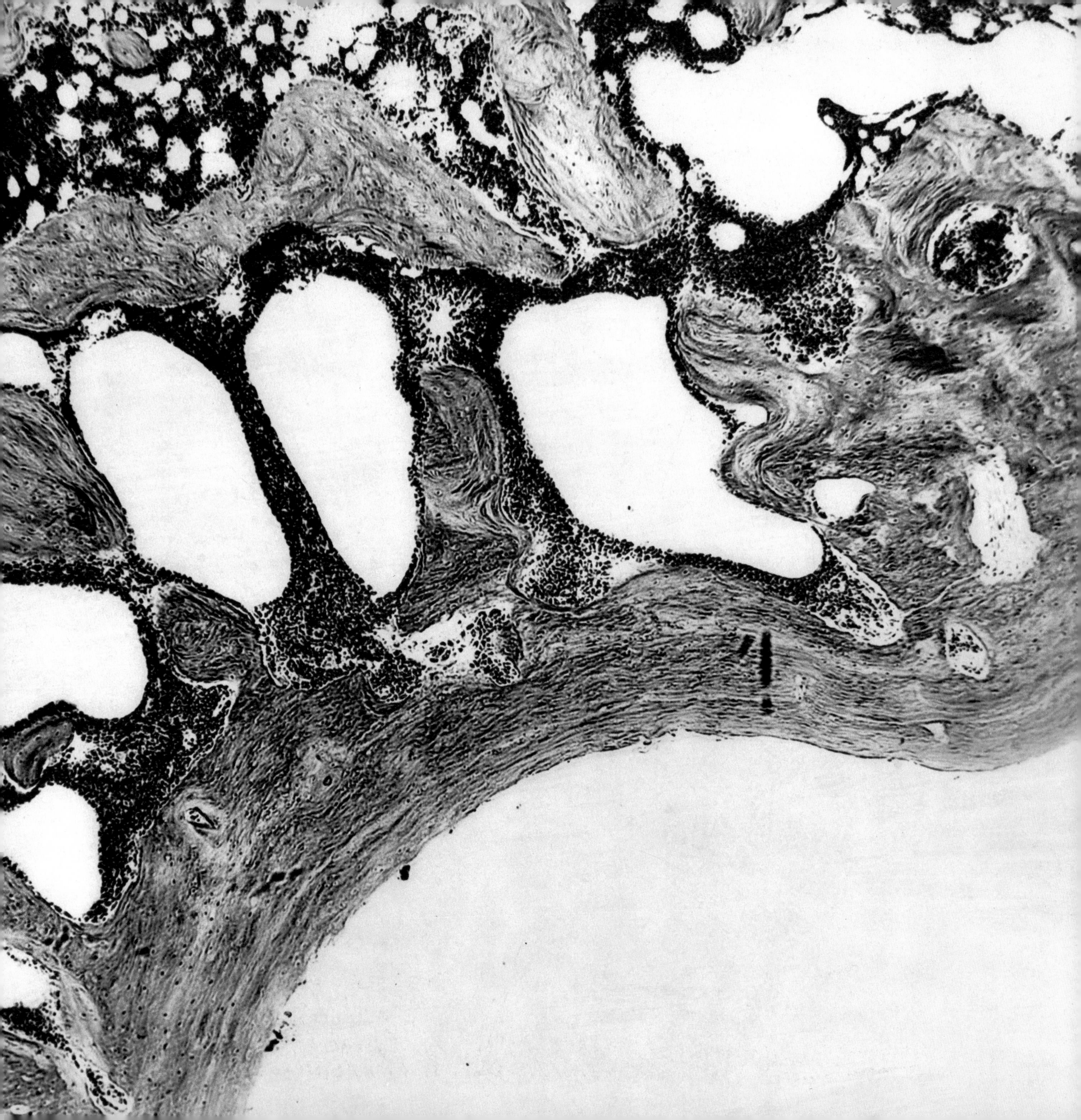

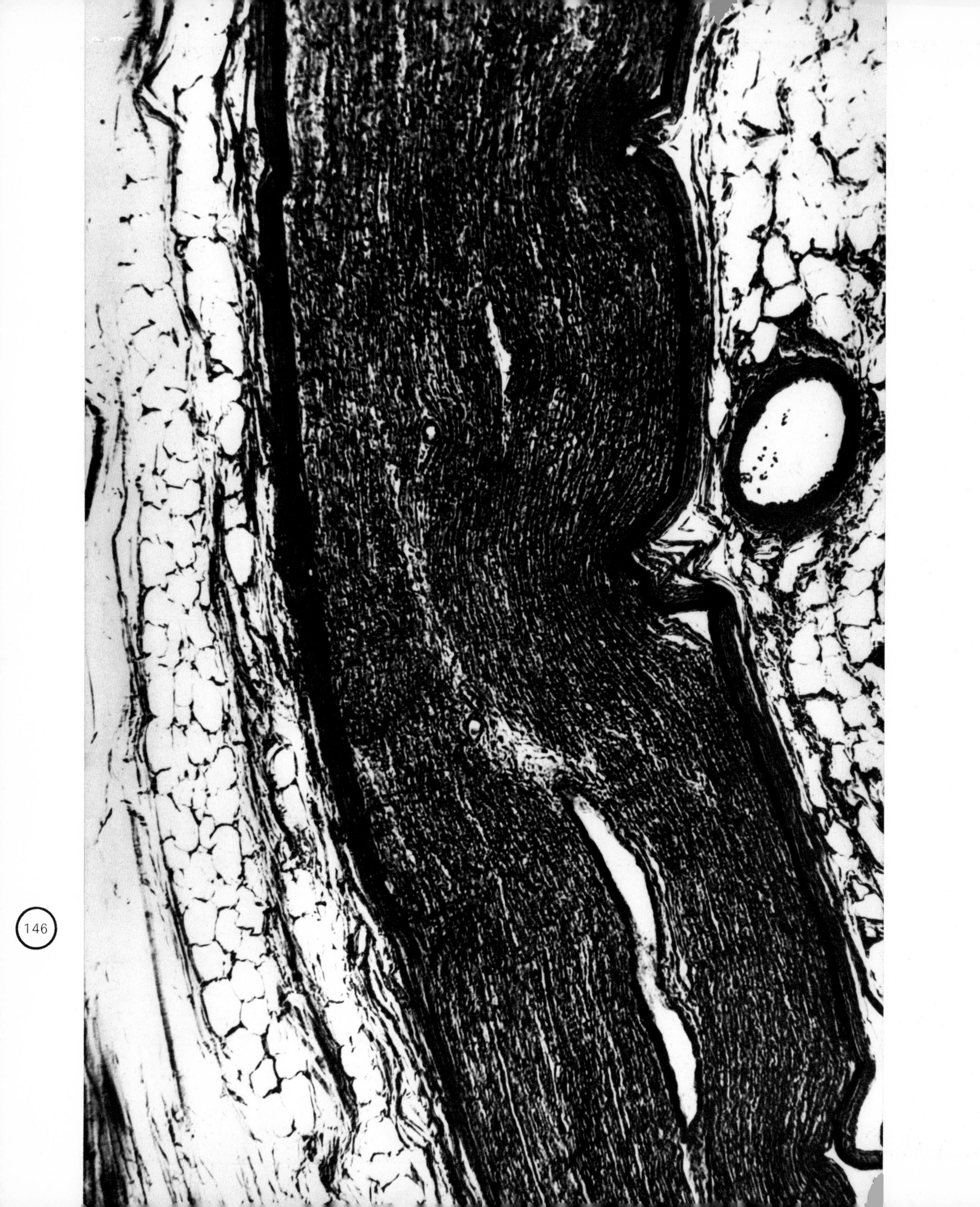

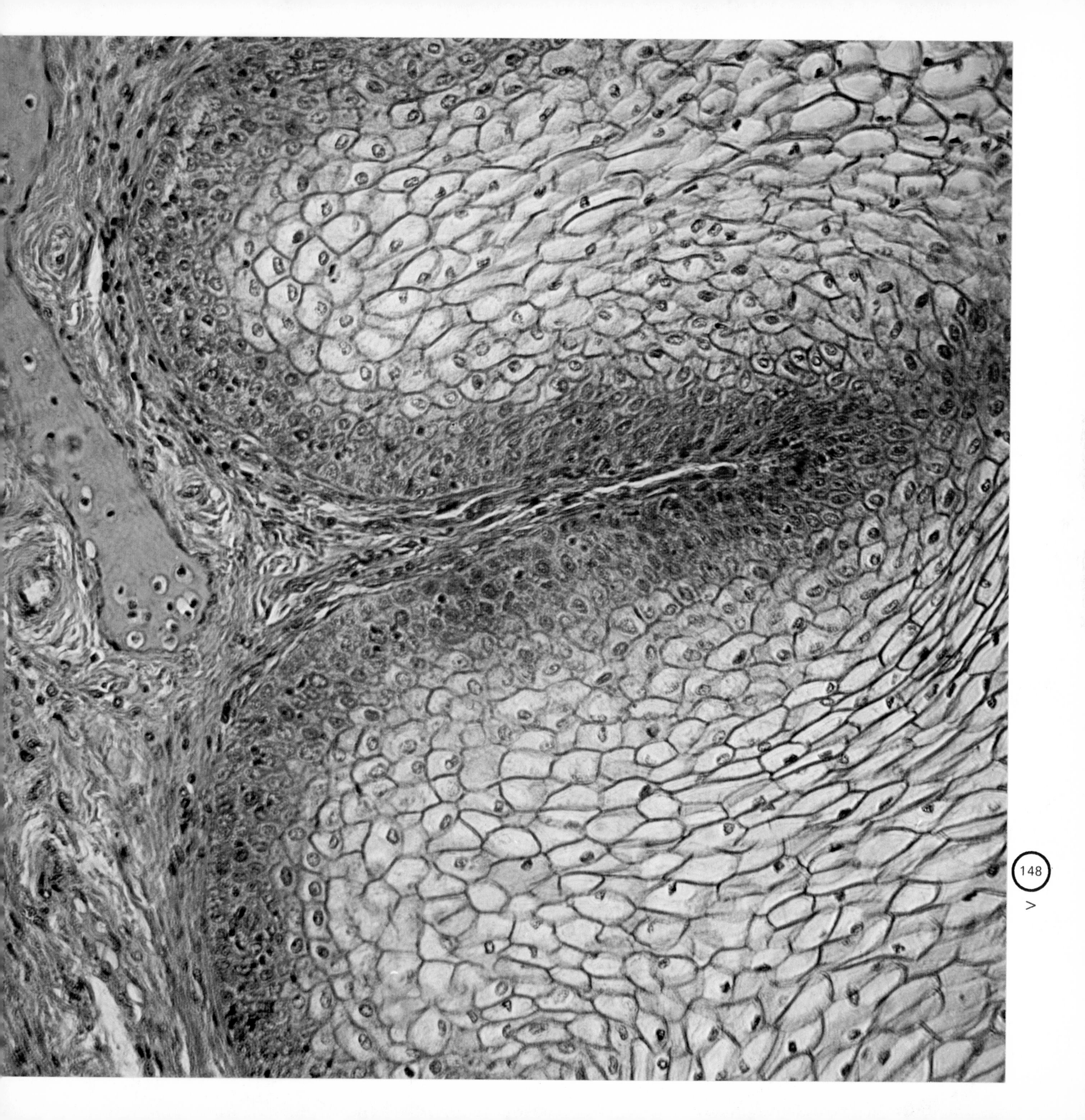
148

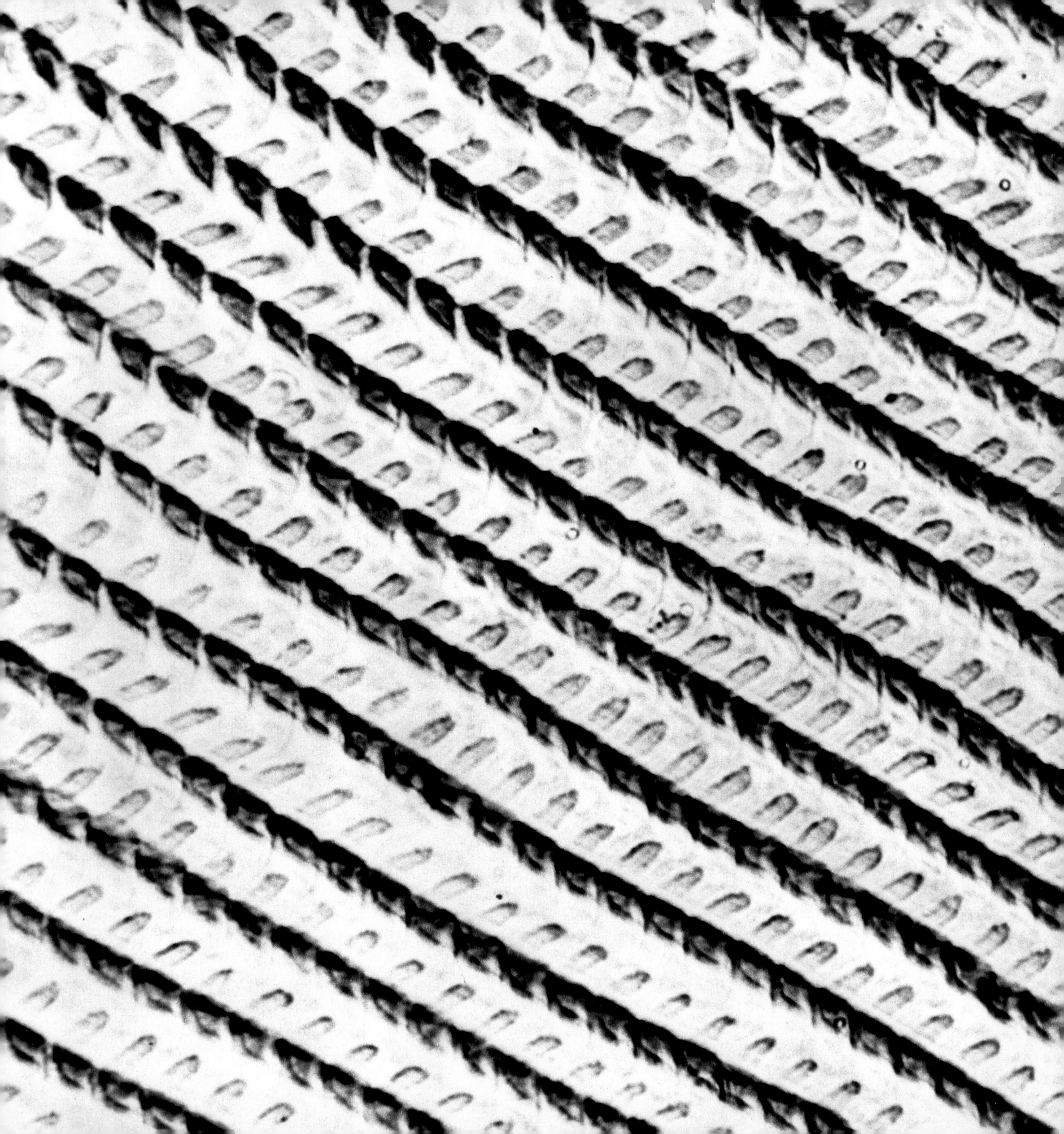

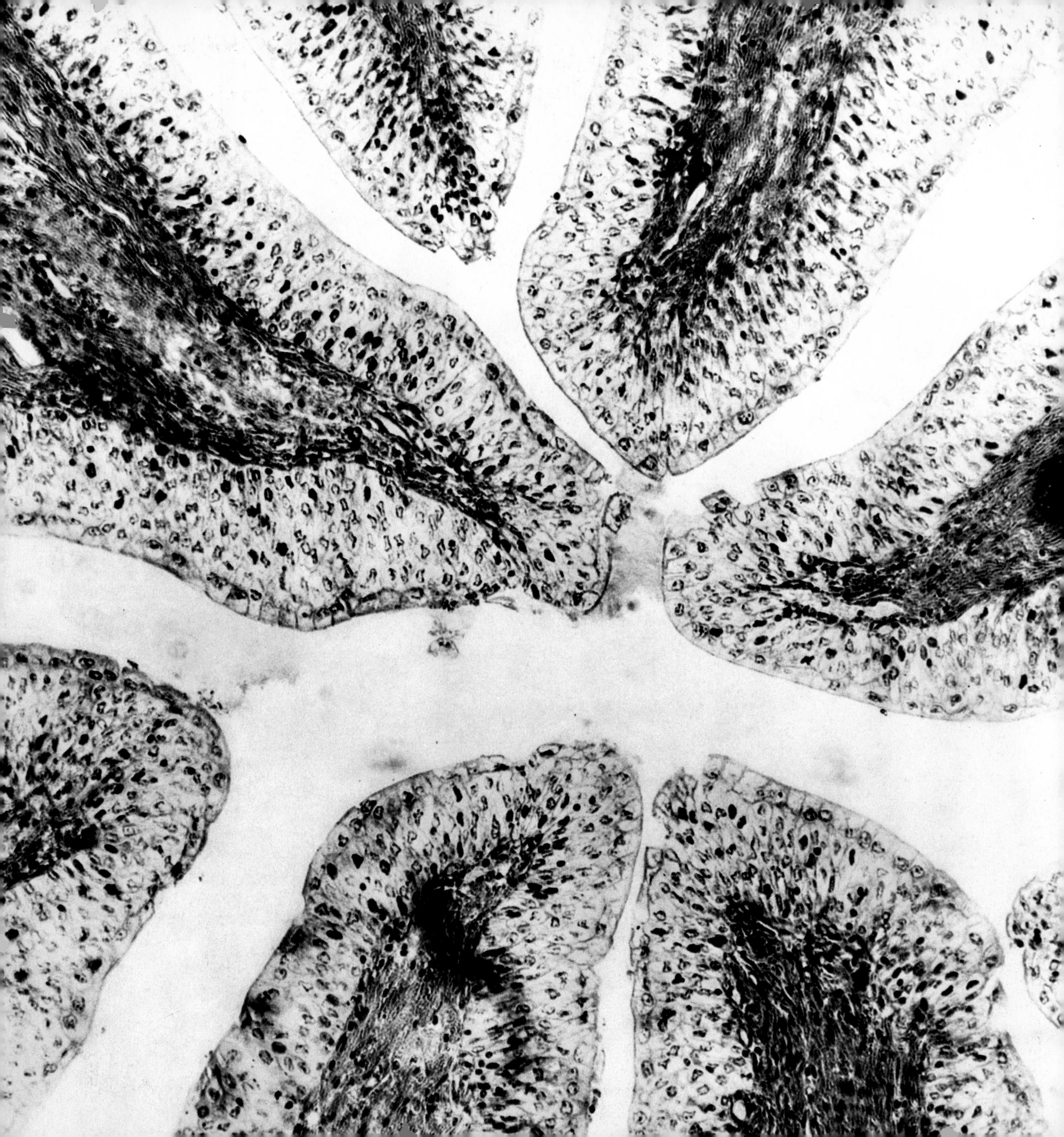

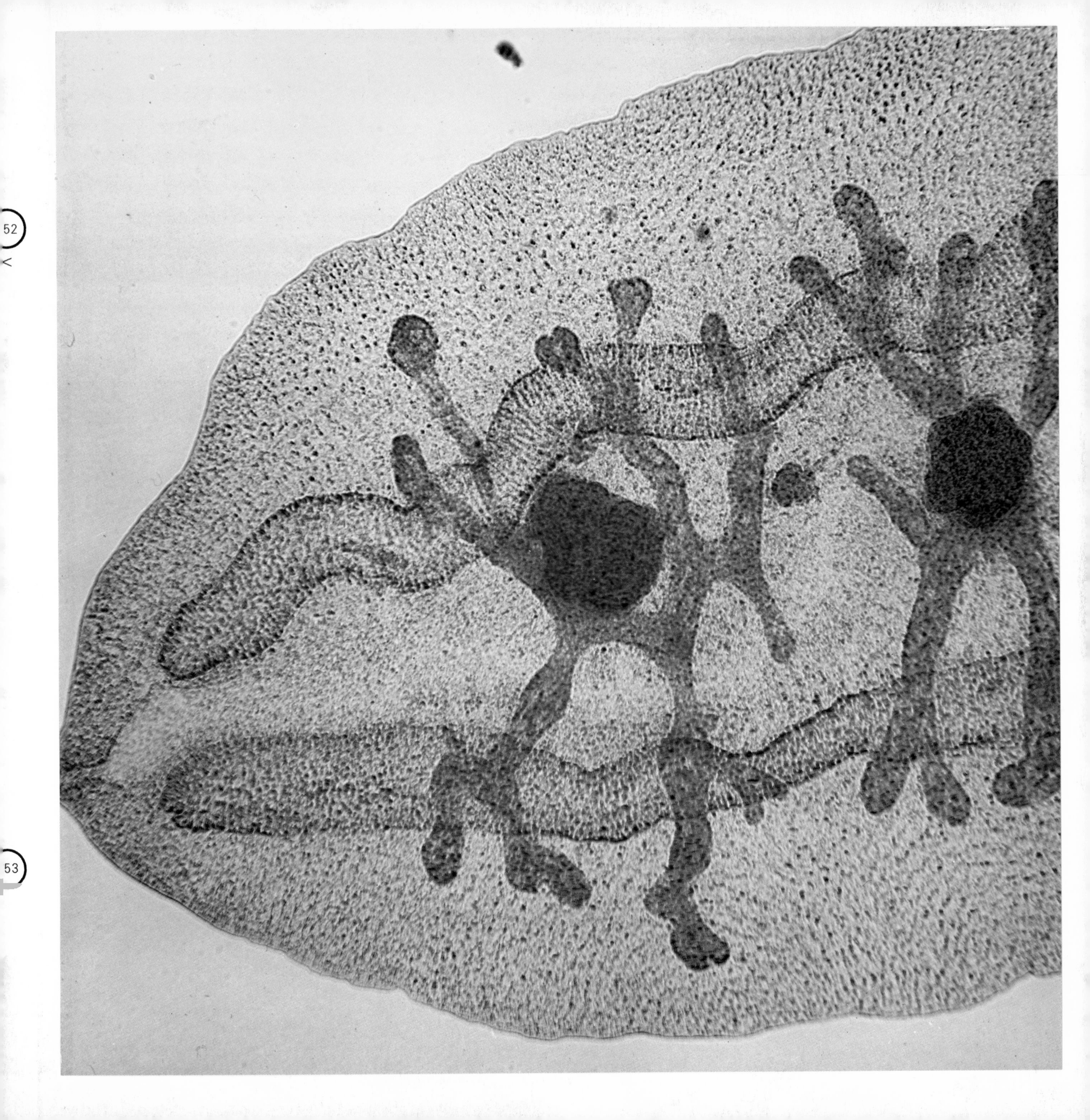

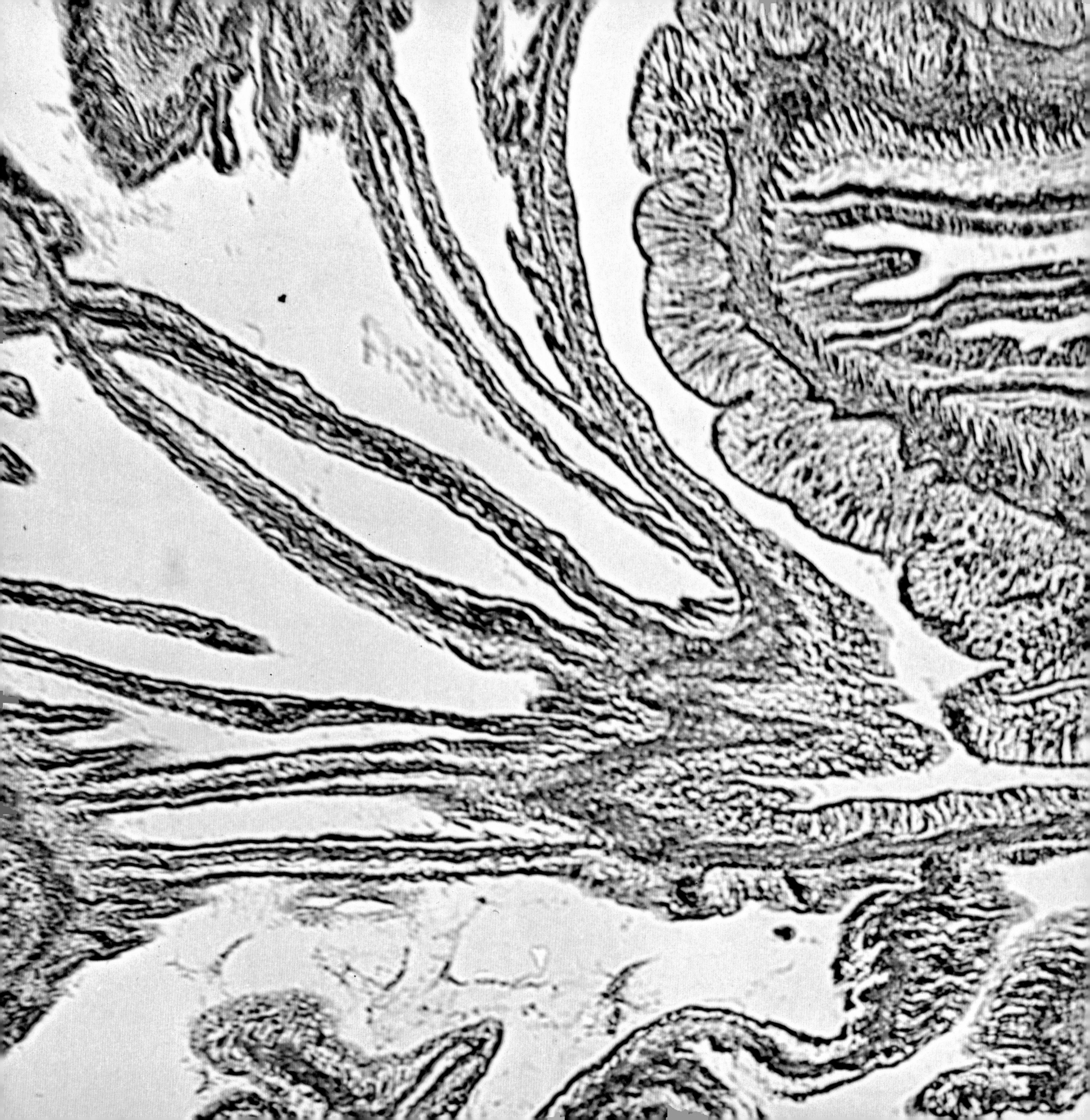

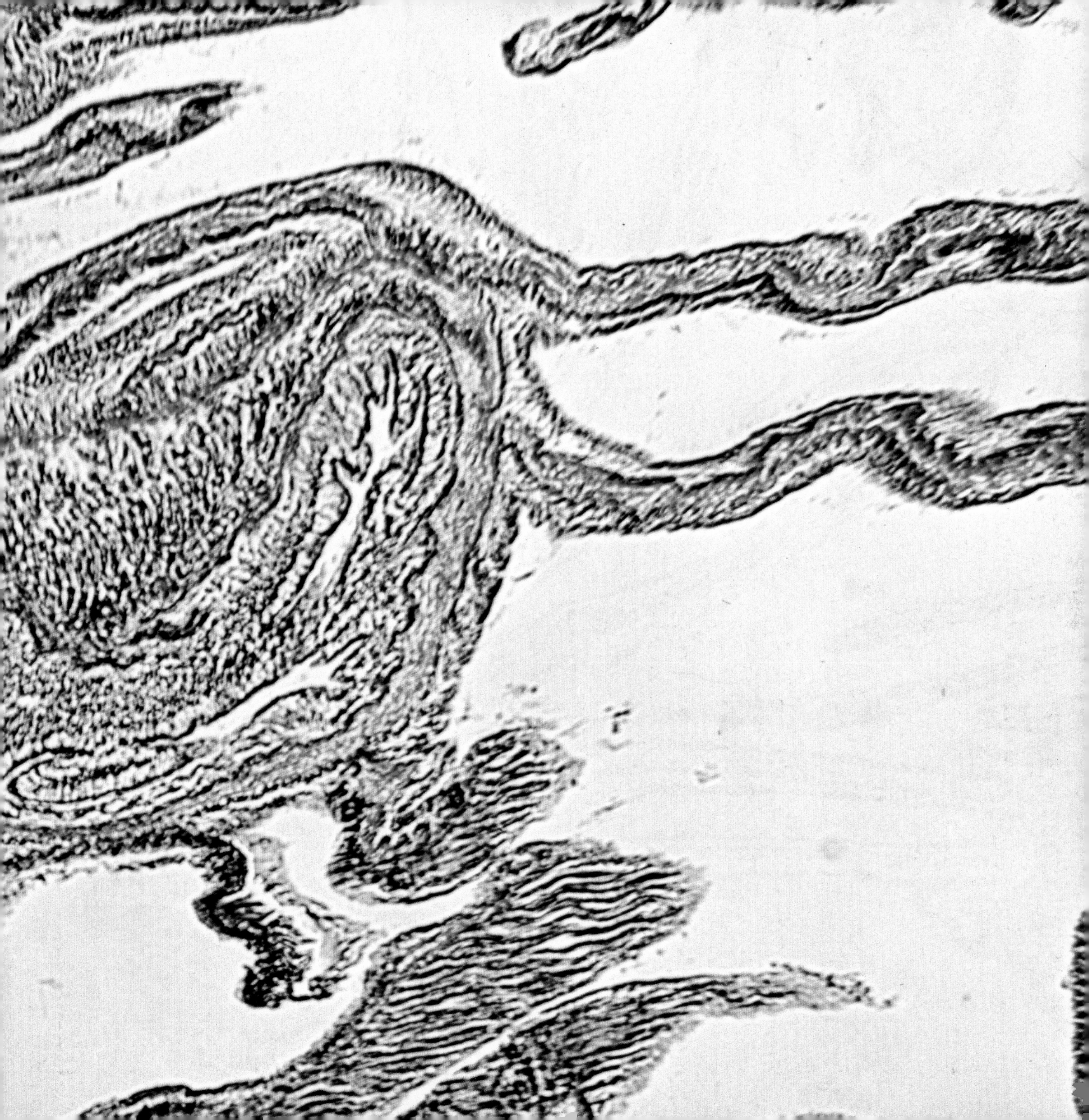

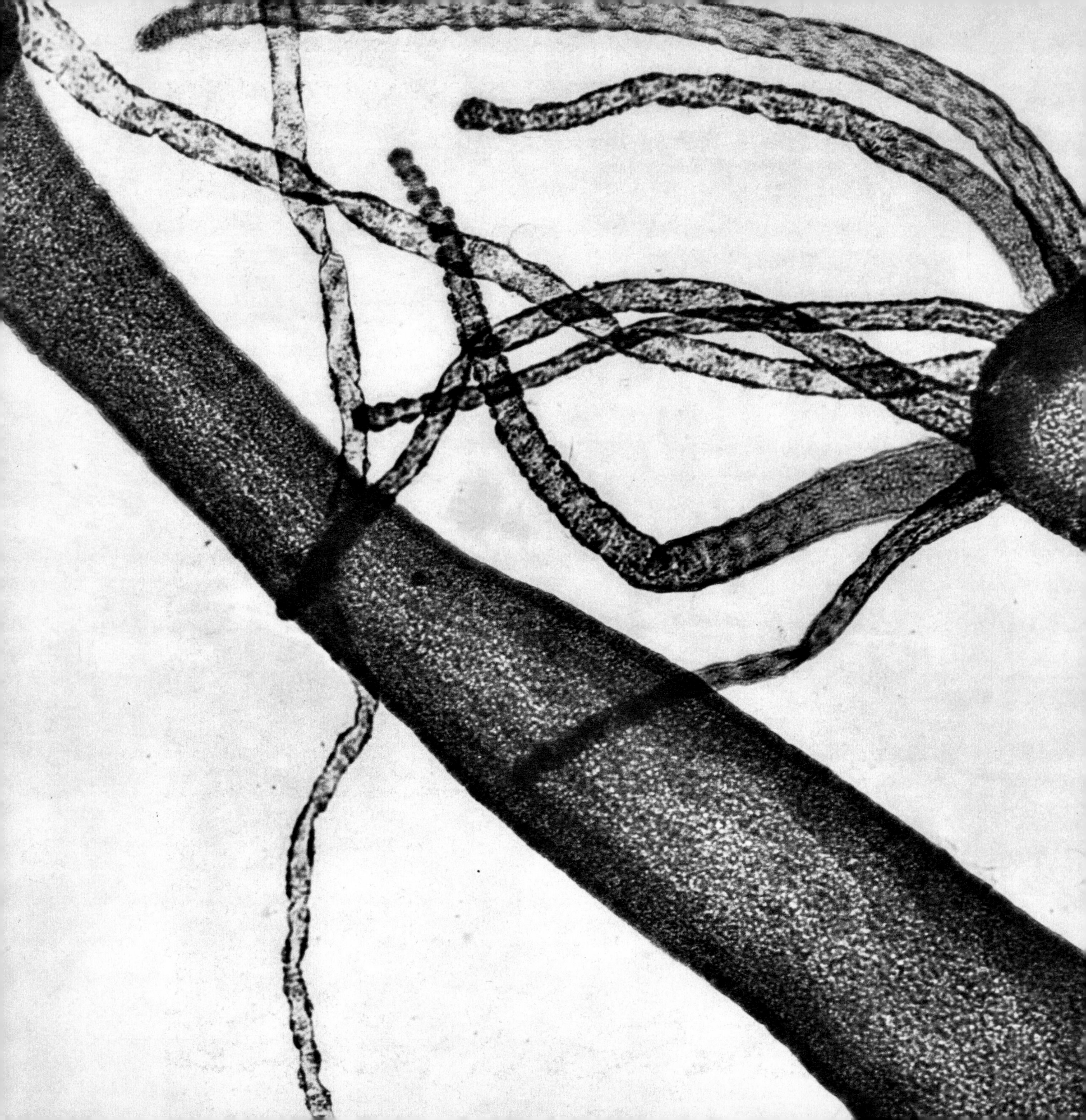

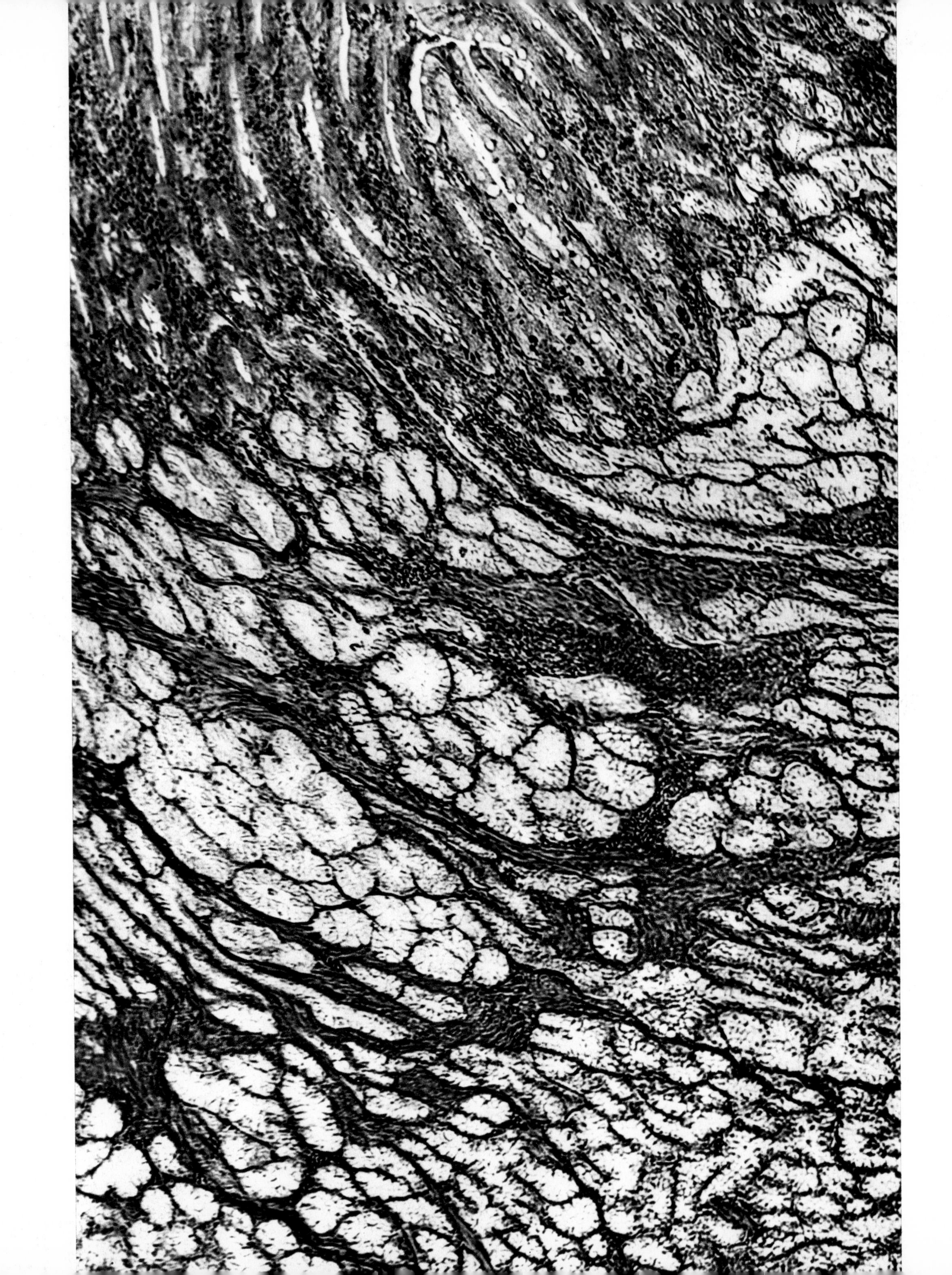

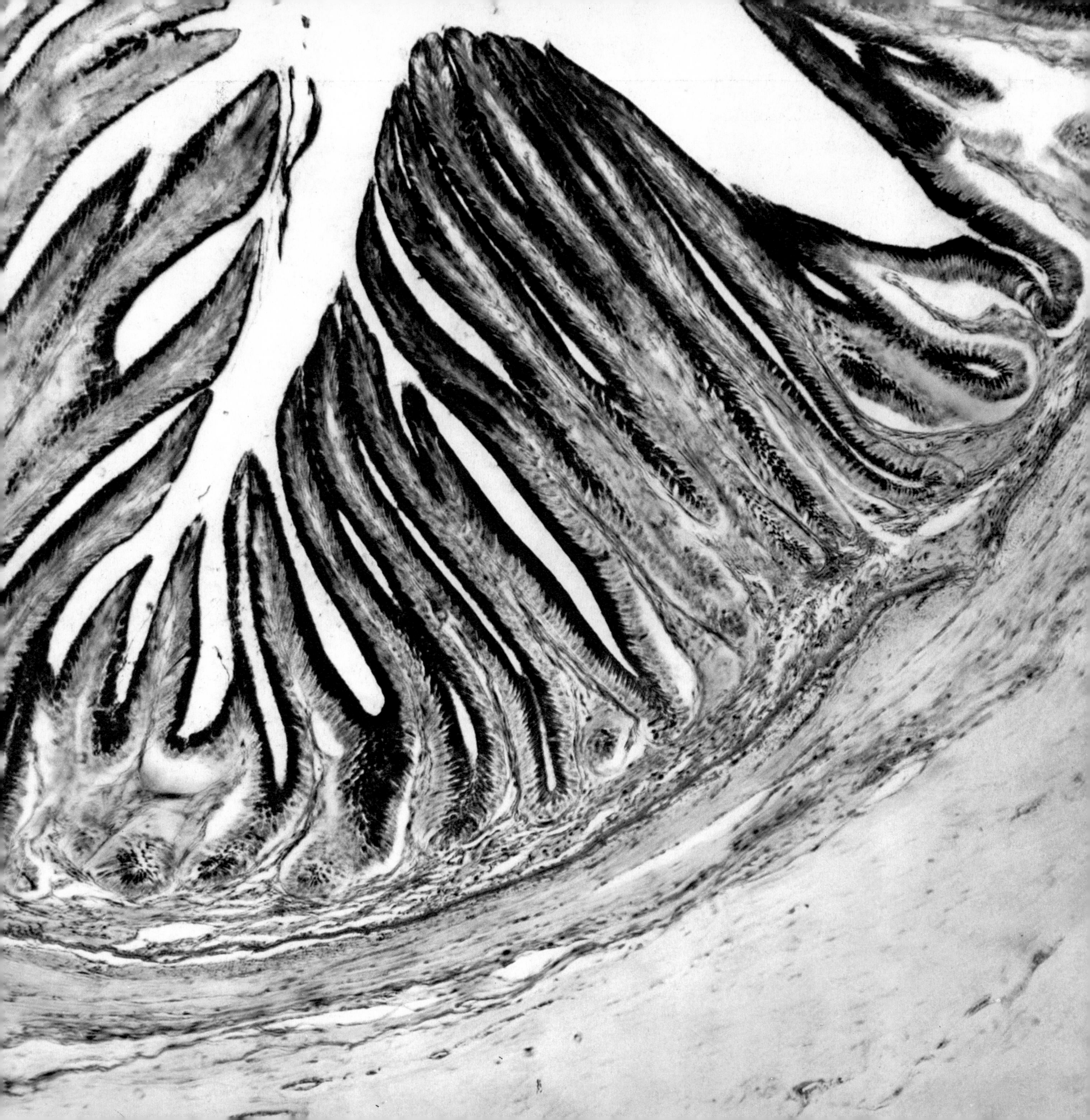

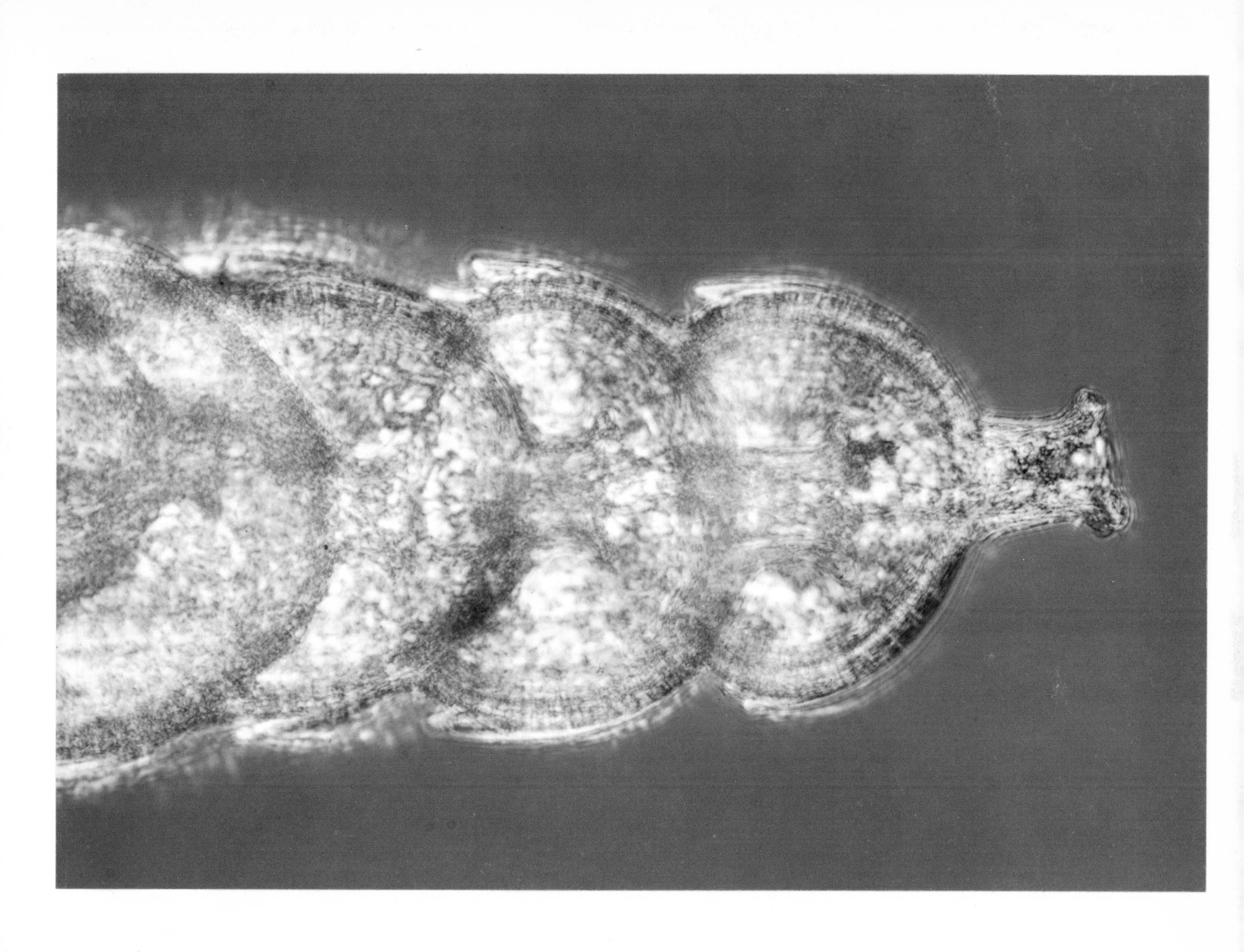

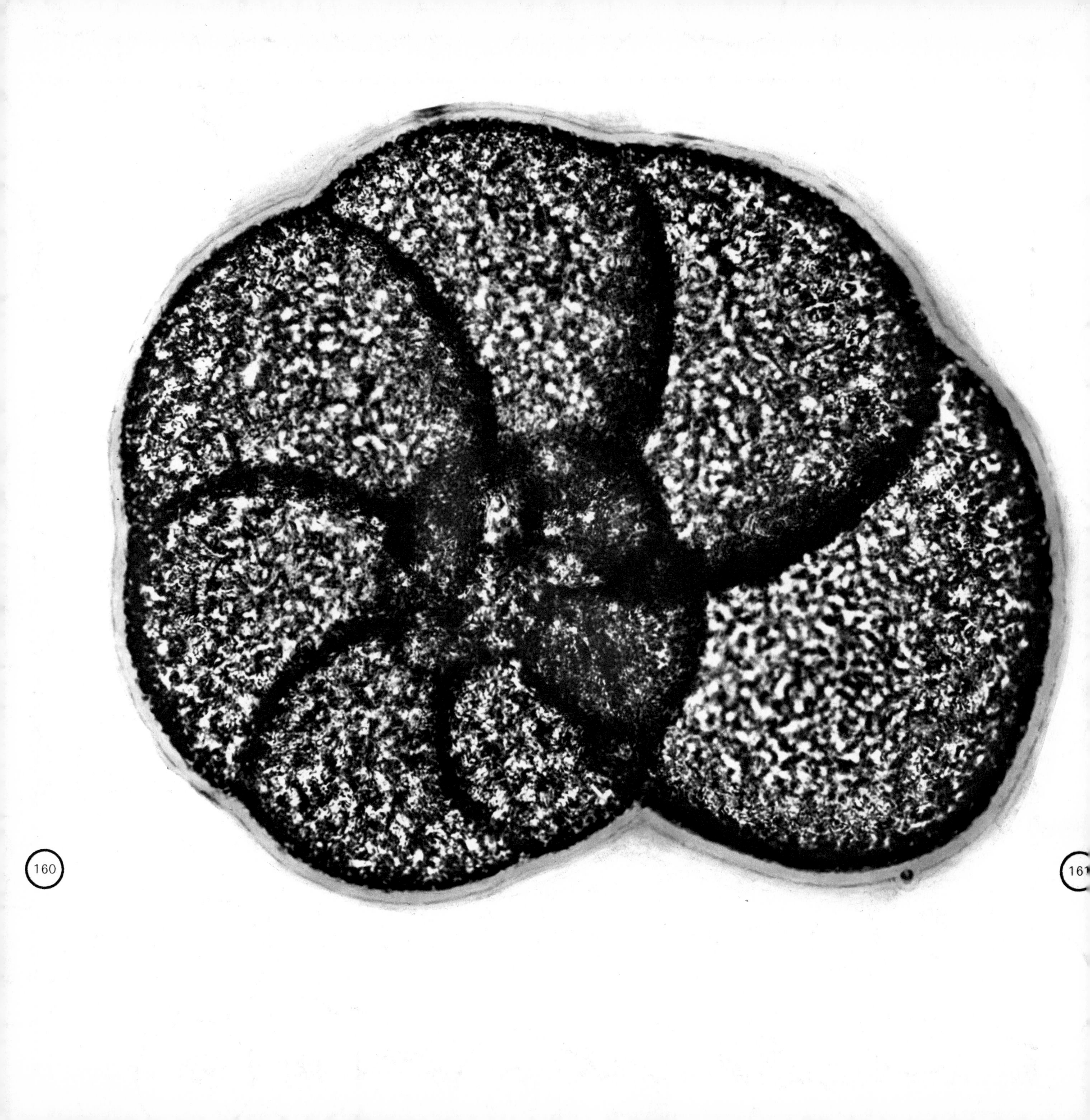
160

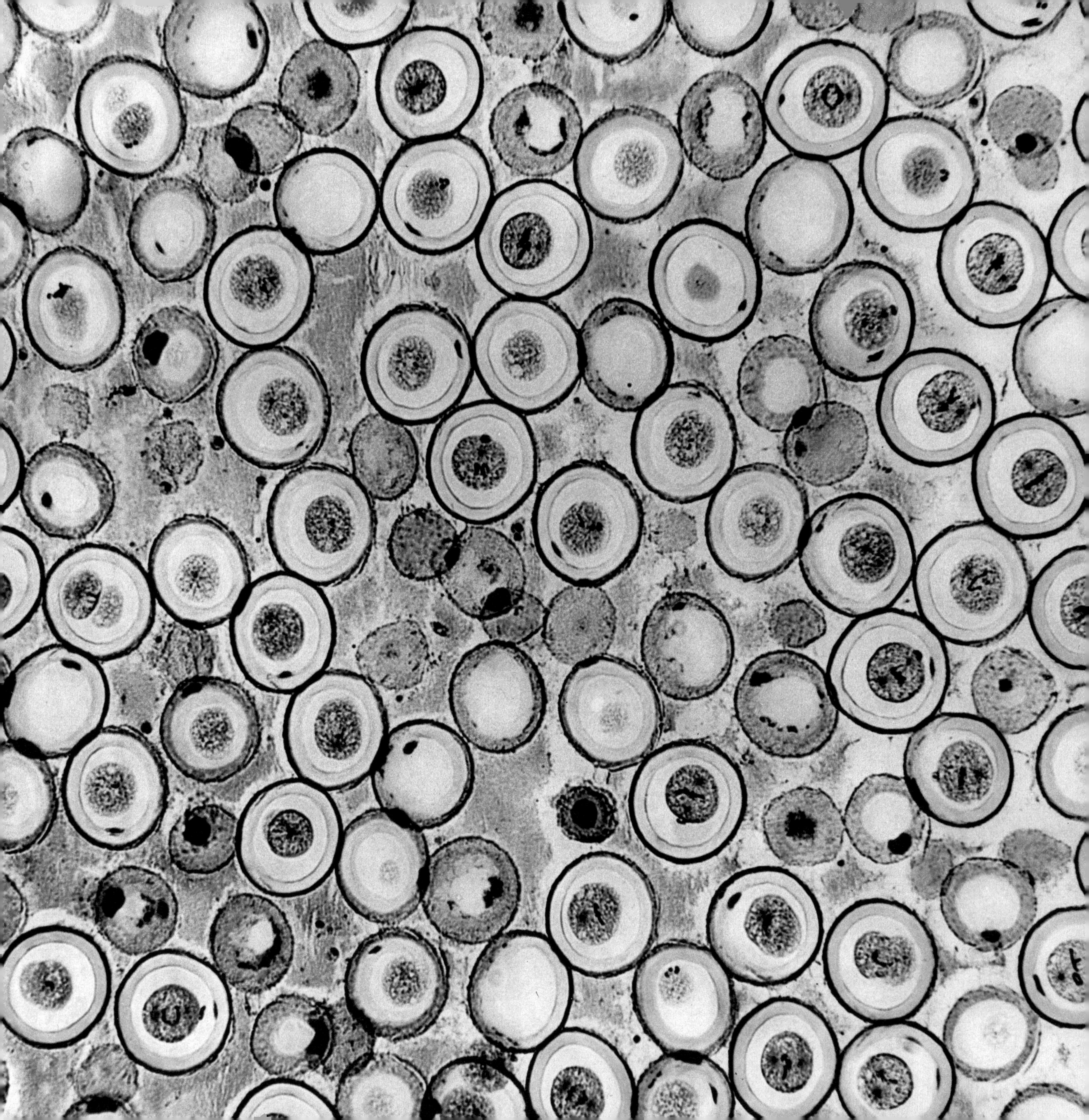

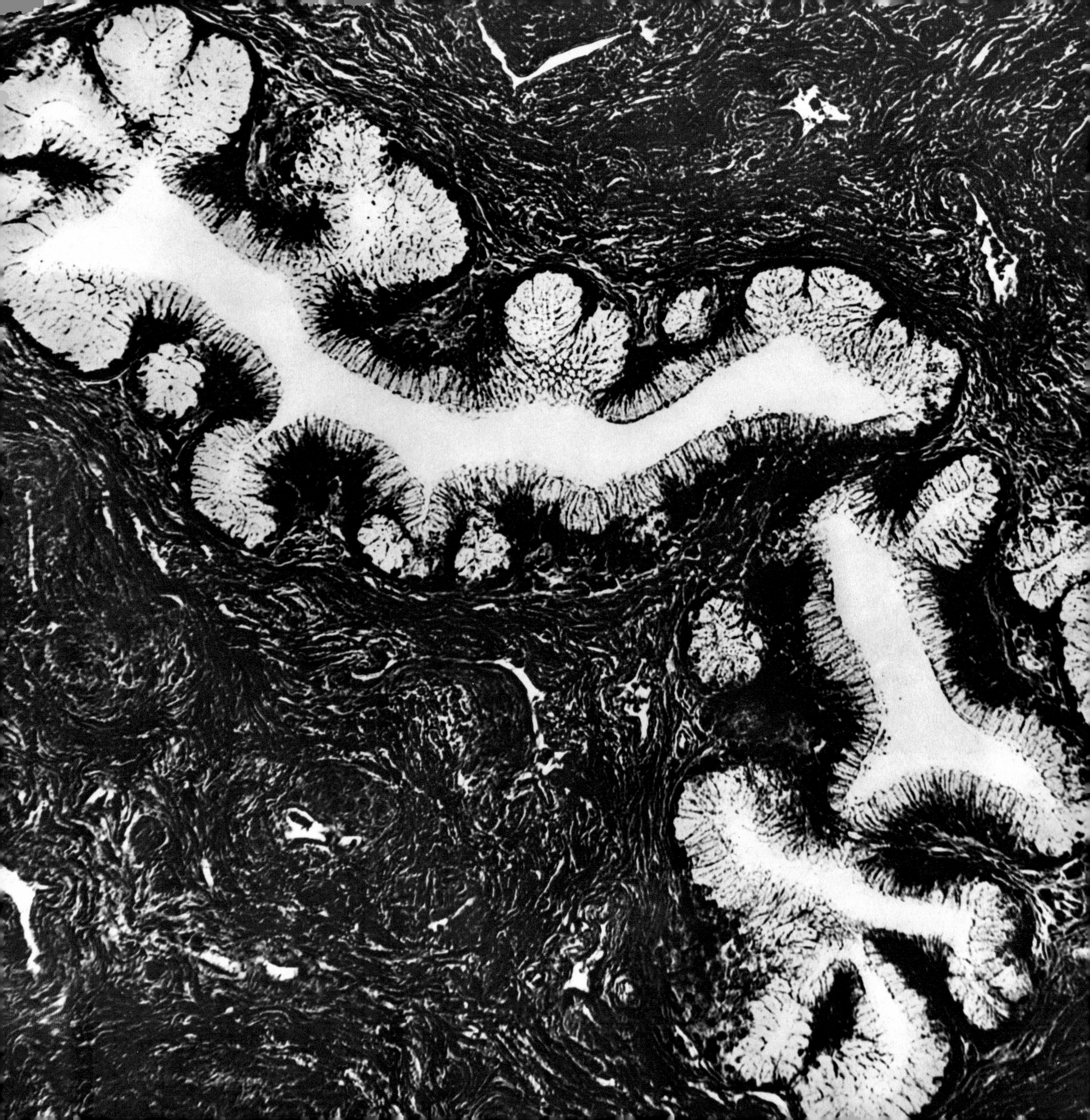

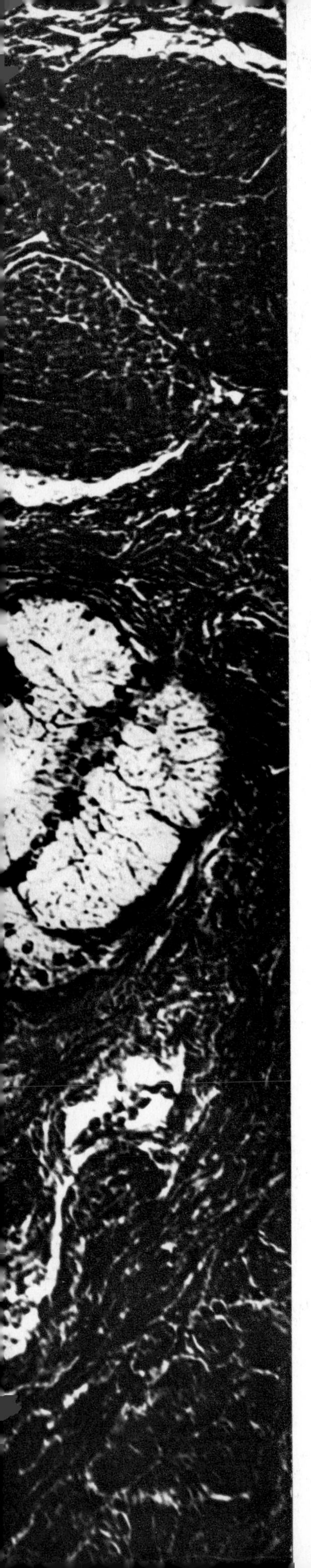

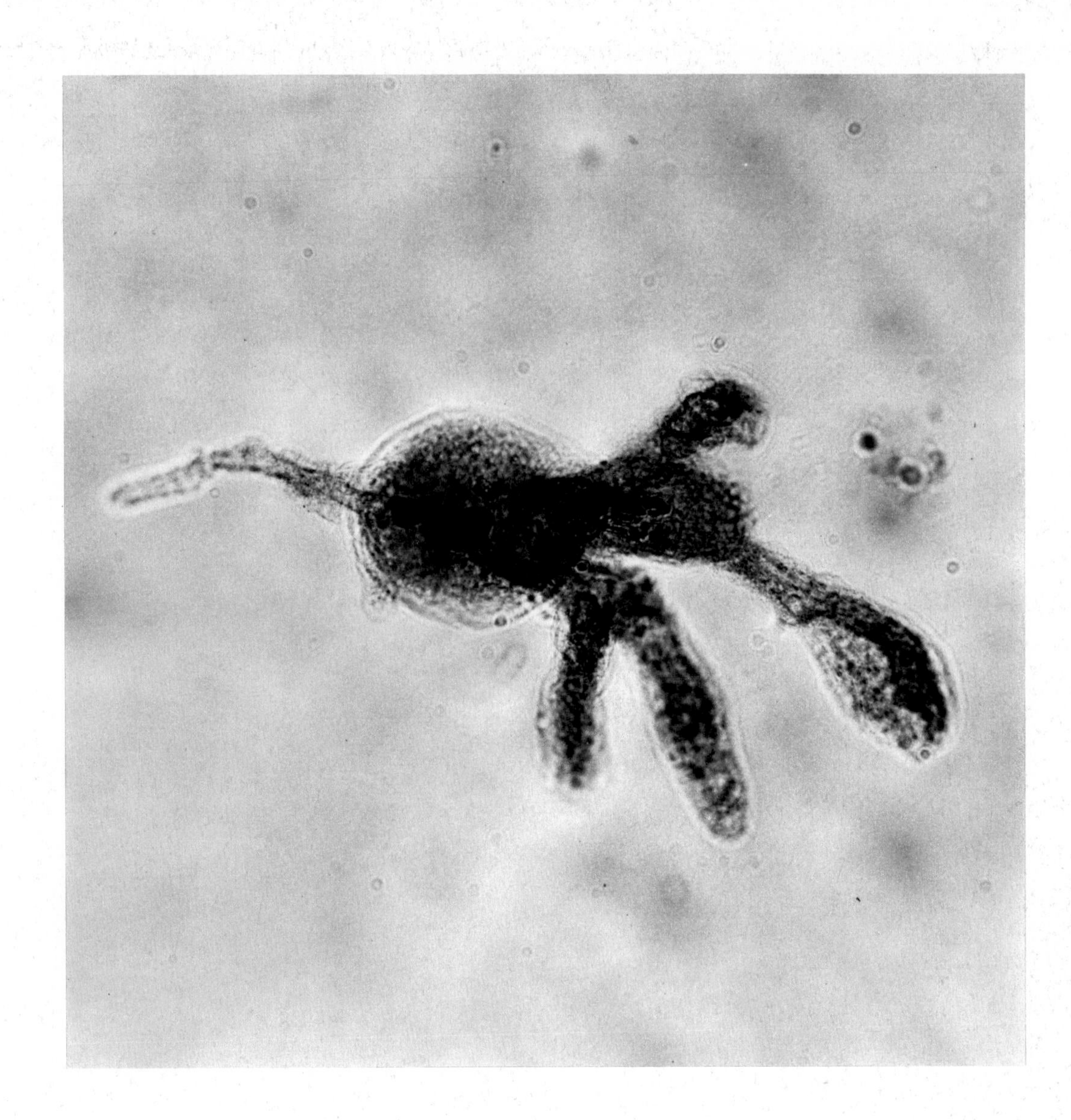

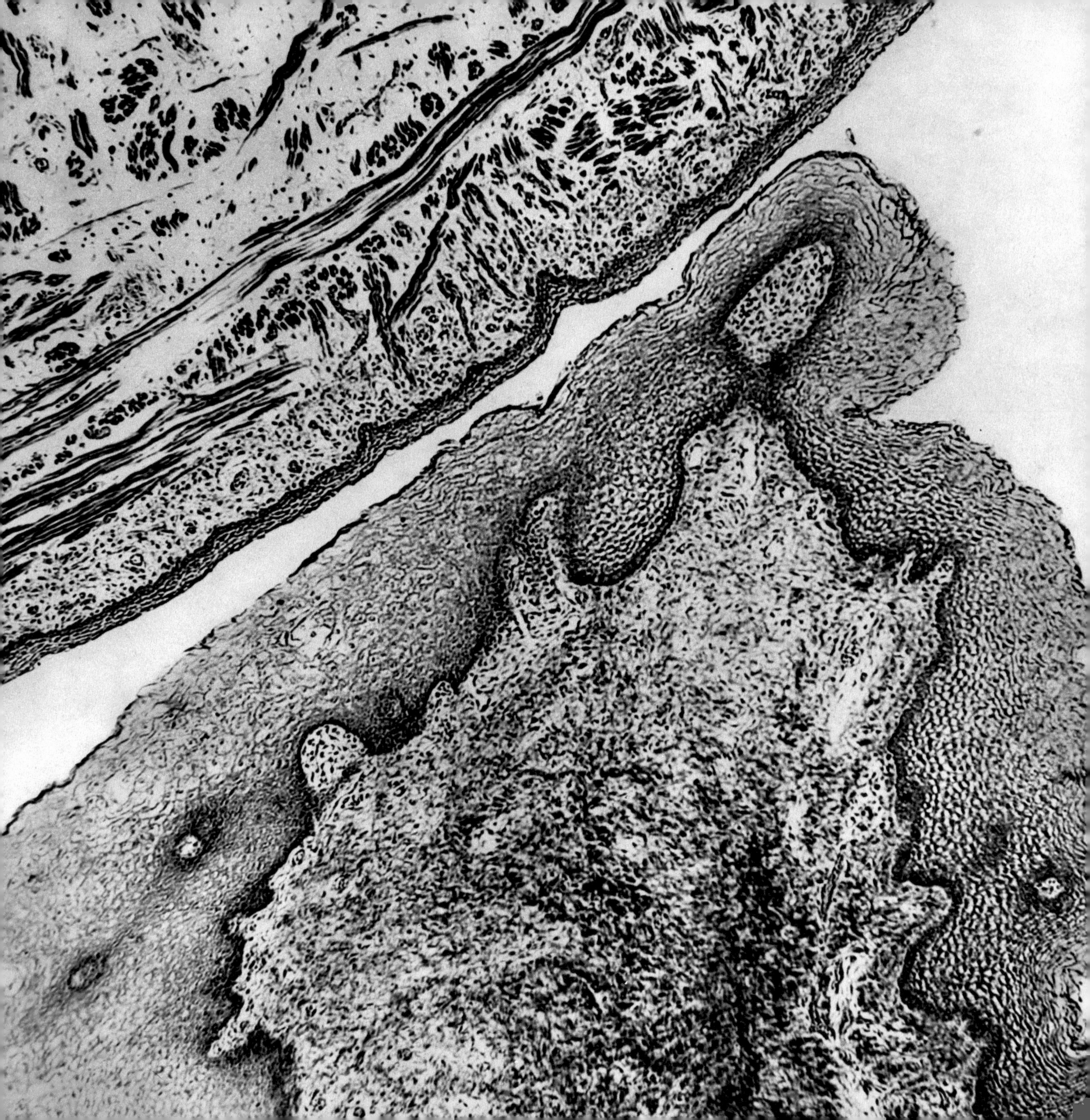

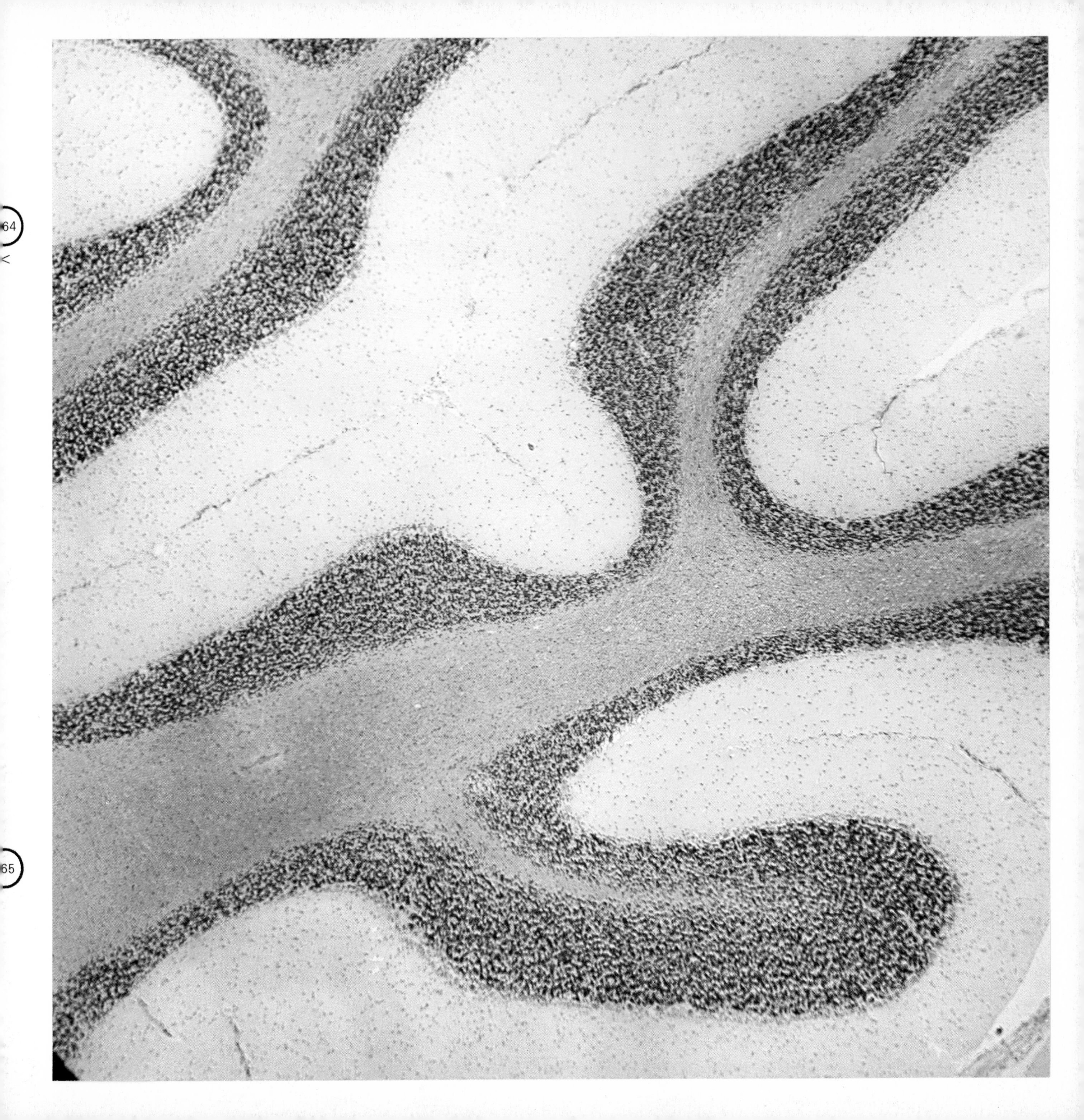

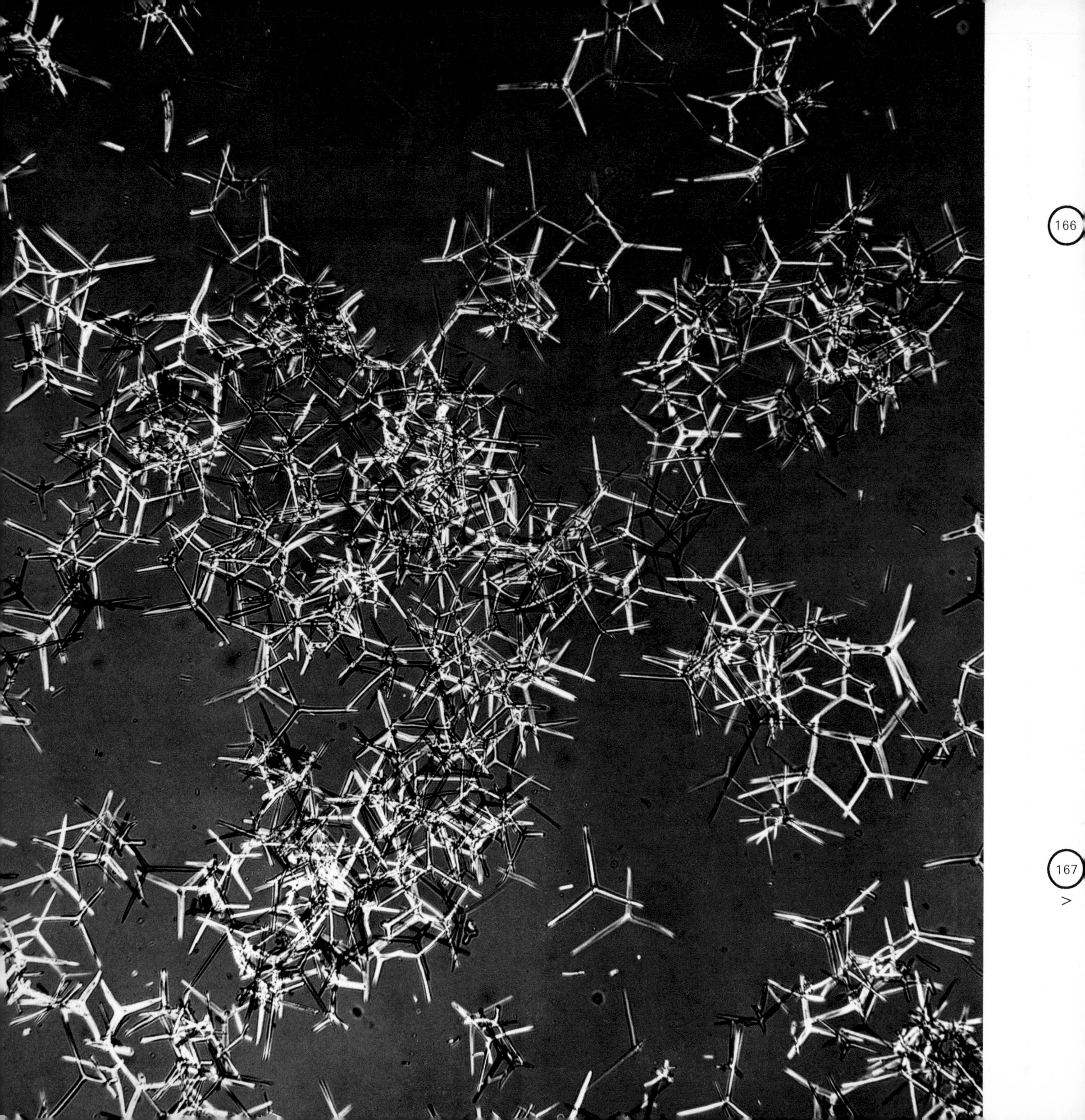

166

167 >

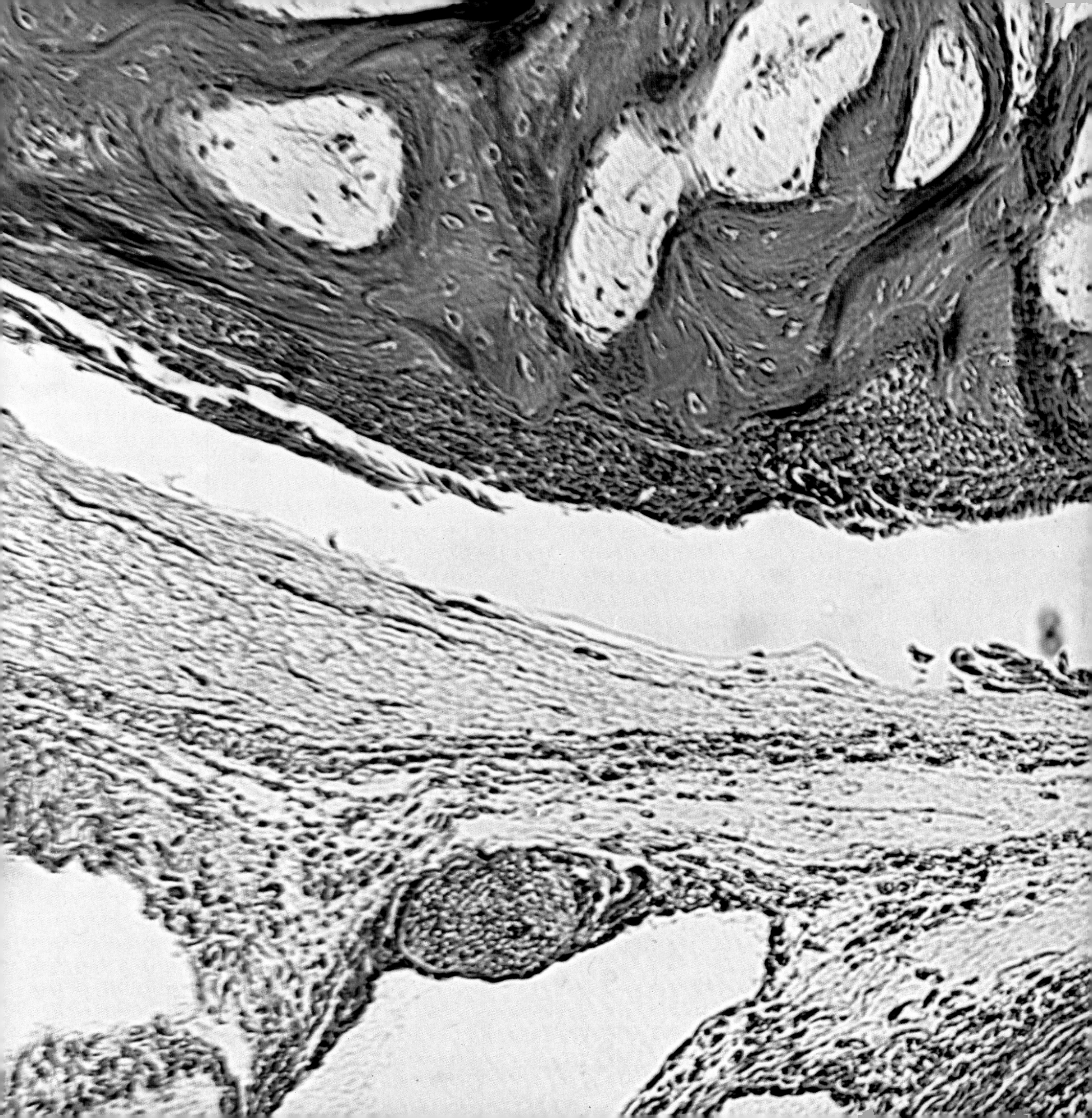

218. **Dendroidal carbide extracted from an intergranular fracture of stainless steel**
125,000x Taken with an electron microscope
Courtesy of Visvaldis Biss, Climax Molybdenum Company of Michigan, Ann Arbor, Mich.

219. **Prehnite**
500x Polarized light

220. **Pyrite, magnolite, hemolite**
150x

221. **Pyrite**
210x

222, 223. **Microstructure of cast 250 grade maraging steel**
30,000x
Courtesy of Dennis S. Crowther, Rudy B. Spanholtz, and Fred Winchell, Martin Marietta Corp., Orlando, Fla.

224. **Silicon wafer**
225x

225. **Rochelle salts (potassium sodium tartrate)**
190x Polarized light

226. **Amino acid: *DL* norleucine**
240x Polarized light

227. **Potassium ferricyanide**
1,200x Polarized light

228. **Etched surface of a fracture in magnesia-alumina specimen**
50,000x Taken with an electron microscope
Courtesy of Clark D. Smith, Space Division, The Boeing Co., Seattle, Wash.

229. **Iron-nickel alloy**
760x Polarized light, stain etching
Courtesy of Arlan D. Benscoter, Bethlehem Steel Corp., Bethlehem, Pa.

230. **Uranium-iron; dendrite along a central axis**
12,400x Taken with an electron microscope
Courtesy of I. D. Miller, General Electric Co., Nuclear Materials and Propulsion Operation, Cincinnati, Ohio.

231. **Orthoclase**
750x Polarized light

left: 232. **Titanium alloy**
Co., Seattle, Wash.
375x
Courtesy of Frank D. Walsh, Manufacturing Research and Development, The Boeing Co., Seattle, Wash.

right: 232. **Titanium alloy**
250x
Courtesy of Frank D. Walsh, Manufacturing Research and Development, The Boeing Co., Seattle, Wash.

left: 233. **Titanium base metal**
435x
Courtesy of Frank D. Walsh, Manufacturing Research and Development, The Boeing Co., Seattle, Wash.

right: 233. **Titanium alloy**
500x
Courtesy of Frank D. Walsh, Manufacturing Research and Development, The Boeing Co., Seattle, Wash.

234. **Hypo (sodium thiosulfate)**
700x Polarized light

235. **Titanium brazed with copper foil**
1,900x
Courtesy of Frank D. Walsh, Manufacturing Research and Development, The Boeing Co., Seattle, Wash.

236. **Amino acid: *DL* tyrosine**
300x Polarized light

237. **Amino acid: glycine**
200x

238, 239. **Gel-type plastic material: liquid crystal (fluor chemical) 3,600x Polarized light**
Courtesy of Carl Zeiss, Inc., New York.

240. **Cholesterol ester**
500x Polarized light

241. **Rochelle salts (potassium sodium tartrate)**
200x Polarized light

242. **Hypo (sodium thiosulfate)**
700x

243. **Potassium ferricyanide**
375x Polarized light

244. **Nickel sulfate**
200x Polarized light

245. **Rusted surface of a broken tool steel part**
125,000x Taken with an electron microscope
Courtesy of Clark D. Smith, Space Division, The Boeing Co., Seattle, Wash.

246, 247. **Martensite (the hardest and most brittle condition of steel) in an iron nickel alloy**
11,600x Polarized light, stain etching
Courtesy of Arlan D. Benscoter, Bethlehem Steel Corp., Bethlehem, Pa.

248. **Albite**
700x Polarized light

249. **Potassium chromium sulfate**
230x Polarized light

250. **Amino acid: *DL* leucine**
750x

251. **Cholesterol ester**
2,000x Polarized light

top: 252. **Copper-silicon alloy**
250x
Courtesy of Buehler Ltd., Evanston, Ill.

bottom: 252. **Brass**
150x
Courtesy of James L. Dvorak, Buehler Ltd., Evanston, Ill.

253. **Quartz**
600x

254. **Hard chromium**
3,600x
Courtesy of Henry A. Kubiak, General Electric Co., Erie, Pa.

left: 255. **Surface of a fake Chinese bronze. 5x**
Photographer, W. T. Chase. Courtesy of Freer Gallery of Art, Smithsonian Institution, Washington, D.C.

right: 255. **Aluminum-copper alloy**
290x
Courtesy of Buehler Ltd., Evanston, Ill.

256. **Amino acid: *DL* leucine**
300x Polarized light

257, 258. **Hypo (sodium thiosulfate)**
1,150x Polarized light

176. Nickel sulfate
300x

177. Tremolite
225x

178. Staurolite
300x

179. Andalusite
240x Polarized light

180. Opal
300x Polarized light

181. Study of graphite
1,300x Interference contrast
Courtesy of Edith M. Raviola, General Electric Co., Research and Development Center, Schenectady, N.Y.

182. Silicon
500x Polarized light

183. Silicon wafer
400x

184, 185. Cubic crystalline structure of etched tungsten
20,600x Taken with an electron microscope
Courtesy of Clark D. Smith, Space Division, The Boeing Co., Seattle, Wash.

186. Cholesterol ester
200x Polarized light

187. Amino acid: glycine
740x Polarized light

188. Detail of a surface coating on titanium metal
4,500x Taken with an electron microscope
Courtesy of Clark D. Smith, Space Division, The Boeing Co., Seattle, Wash.

189. Sugar
200x Polarized light

190. Cholesterol ester
500x Polarized light

191. Cholesterol ester
375x Polarized light

192, 193. Maraging steel
9,000x
Courtesy of Frank D. Walsh, Manufacturing Research and Development, The Boeing Co., Seattle, Wash

Magnesium alloy
Courtesy of Frank D. Walsh, Manufacturing Research and Development, The Boeing Co., Seattle, Wash.

194. Nickel-aluminum bronze, transverse cracks during tensile test 32,000x
Courtesy of W. Gifkins, British Welding Research Association, Cambridge, England.

195. Amino acid: *DL* norleucine
200x Polarized light

196. Cast steel (pearlite)
100,000x Taken with an electron microscope
Courtesy of Ulla Fuchs and Ursula L. Schaaber, Institut für Harterei-Technik, Bremen, Germany.

197. Field ion micrograph of a nickel molybdenum alloy to study composition at an atomic level
3,000,000x
Courtesy of R. W. Newman, B. G. LeFevre, J. J. Hren, University of Florida, Gainesville, Fla.

198. Dayton iron meteorite
475x Polarized light with interference plate
Courtesy of Harvey Yakowitz, National Bureau of Standards, Wash. D.C., and J. I. Goldstein, Goddard Space Flight Center Greenbelt, Md.

199. Amino acid: *DL* leucine
800x Polarized light

200. High carbon steel (martensite plate)
25,000x Taken with an electron microscope
Courtesy of Arlan D. Benscoter, Bethlehem Steel Corp., Bethlehem, Pa.

201. Copper-cadmium-zinc alloy
16,000x
Courtesy of Karin Baer, Max Planck-Institut für Metallforschung, Stuttgart, Germany.

202. Epsom salts
700x Polarized light

203. Resorcinol doped with tartaric acid: variations in alignment of crystals
60x Polarized light
Courtesy of Andrew S. Holik, General Electric Co., Research and Development Center, Schenectady, N.Y.

204. Vesuvianite
300x

205, 206. Amino acid: *DL* isoleucine
490x Polarized light

207, 208. Chromium-nickel subjected to sulfidation
14,900x
Courtesy of Robert R. Russell, Research and Development Center General Electric Co., Schenectady, N.Y.

209. Rochelle salts (potassium sodium tartrate)
200x Polarized light

210. Noibium (columbium) subjected to vacuum degassing at 2,300 °C under strong pressure
8,800x Taken with an electron microscope
Courtesy of Robert W. Meyerhoff, Union Carbide Corp., Indianapolis, Ind.

211. Nepheline
180x Polarized light

212, 213. Clont (1-hydroxy ethyl-2-methyl-5-nitroimidazo
575x Polarized light
Courtesy of Carl Zeiss, Inc., New York.

214. Crystal growth in the oxidation of lead
42,600x Taken with an electron microscope
Courtesy of Edward F. Koch, General Electric Co., Schenectady, N.Y

***top:* 215. Copper-zinc alloy**
500x
Courtesy of Buehler Ltd., Evanston, Ill.

***bottom:* 215. Molybdenum-titanium alloy**
250x
Courtesy of Buehler Ltd., Evanston, Ill.

216. Graphite subjected to tension to induce a fatigue fracture
175,000x Taken with an electron microscope
Courtesy of Floyd B. Brate, Jr., General Electric Co., Aircraft Engine Group, Evandale, Ohio.

217. Titanium sheet subjected to nitrogen at a high temperature
600x Polarized light with interference plate
Courtesy of Harvey Yakowitz, National Bureau of Standards, Wash. D.C.

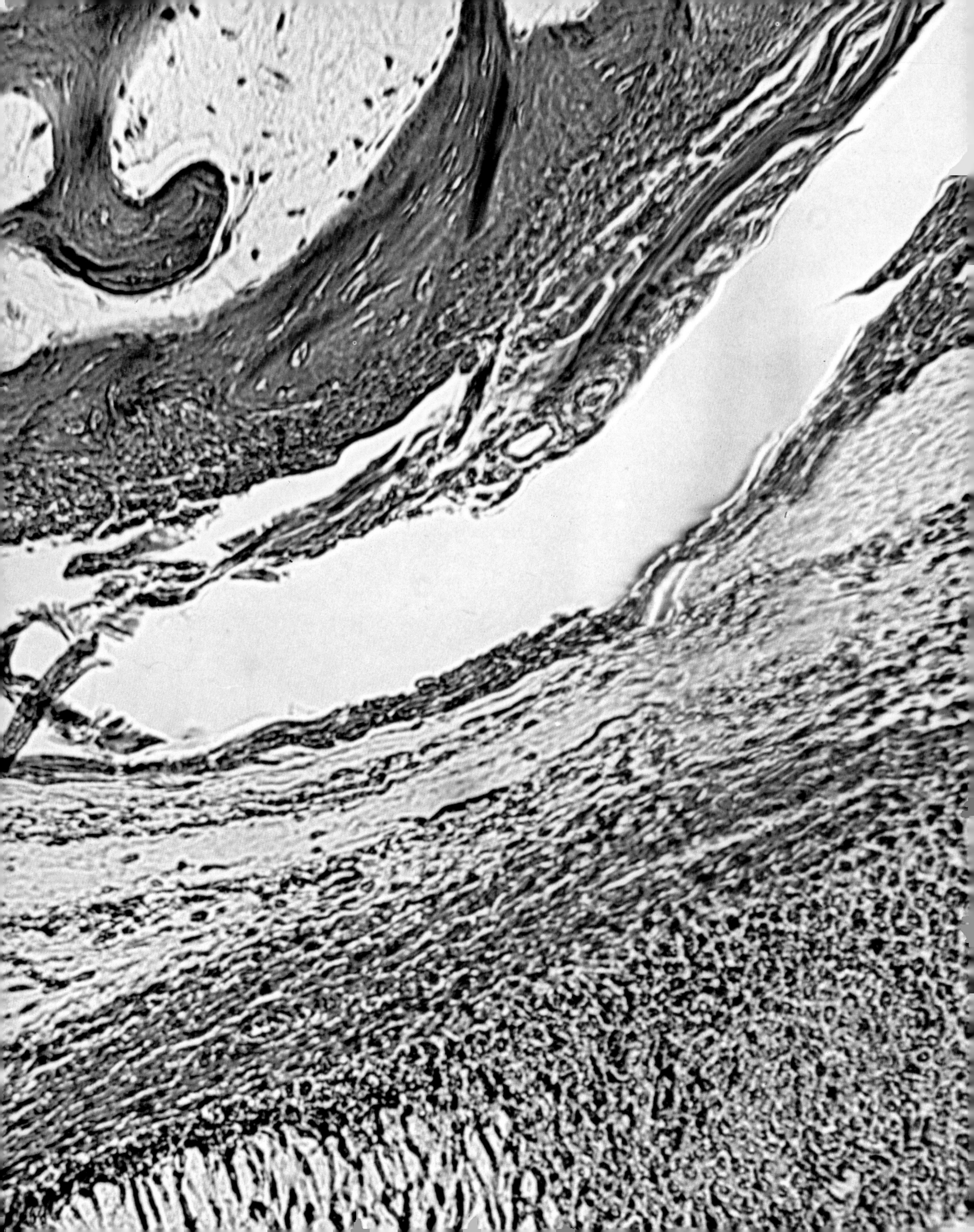

MINERAL

Keep fold-out (at left) open for captions to the following section

Customarily, anything that is not animal or vegetable is considered to be "mineral." This classification is a misleading one, since it contains many nonmineral constituents, including crystals, metals, and a great variety of chemicals. The artificial fusion of different metals into alloys and the combination of chemical substances into new compounds have produced countless types of materials useful to man in his many life functions. The microscopic forms encountered are almost infinite in their diversity. The samples illustrated in this section are selected on the basis of their visual beauty and variety. For convenience, three classes of nonliving substances are included: crystals, metals, and minerals in pure form.

Crystals are homogeneous substances bounded by flat surfaces in various geometrical shapes. The directional forces within crystals determine their configuration and are related to the existing atomic arrangement, which is unique and stable for each crystal class. The angles between corresponding faces of two crystals, irrespective of size, are always the same for the same substance. Crystals accordingly repeat their own spatial structure. Minerals, metals, and the great bulk of chemicals are crystalline in nature.

Crystallography, the science of crystals and the crystalline state, is said to have been founded in 1669 by a Danish physician, Nicolaus Steno, who observed that quartz crystals—no matter how different their appearance—are structured in such a way that the angles between their corresponding faces are always the same. In 1772, J. B. L. Romé de Lisle, confirming Steno's findings, demonstrated that each crystalline form has uniquely characteristic interfacial angles. With the development of X-ray diffraction techniques, the molecular structure of crystals could be more accurately defined, and it became possible to determine how atoms are arranged in crystals, what their sizes are, the distances from one crystal to another, and the internal patterns of all types of crystals. Crystals were found to consist of atoms, molecules, or ions arranged in perfect geometrical patterns.

Practically all common crystals fall into one of twelve structural classes. In visual terms, there are six basic and distinctive styles of crystal architecture: *cubic* (garnet, gold, iron), where all angles are right angles and sides are all equal; *tetragonal* (zircon, tin), with right angles but two varying side lengths; *orthorhombic* (topaz, sulfur), with right angles but three different side lengths; *monoclinic* (gypsum, cane sugar), which differs from orthorhombic in that eight of the angles are not right angles; *triclinic* (microcline, potassium dichromate), which contains no right angles and three different side lengths; and *hexagonal* (emerald, zinc), which have right angles at their vertical sides and top and bottom faces. Pure crystals have a known internal order based on their atomic structure. This inherent order, however, may be disturbed by impurities in the substance and by contaminations with other crystalline materials—creating an endless variety of visual forms and patterns.

In crystal research the microscope has contributed to scientific discoveries in the fields of inorganic chemistry, mineral chemistry, geochemistry, and metallurgy, by helping to define the simple crystal compositions of chemical, mineral, and metallic materials.

The photomicrographs of crystals were taken with the use of two crossed polarizers, rotated to produce a dark field. This apparatus enabled the crystals to split the light and illuminate themselves brilliantly in various colors. As the crystal specimens are shifted around within the polarized light beam, their colors change in bursts of startling beauty.

Metals

Metals are complex aggregates of small crystals packed tightly together in specific structural arrangements. Through the techniques of *metallography* (the photomicrography of metals) scientists have obtained data about the physical make-up of metals, which has contributed greatly to the science of metallurgy.

Metallurgy began as an ancient craft, dating back many centuries before the birth of Christ. The Egyptians, Sumerians, Minoans, Etruscans, and others used a variety of metallic substances for both decorative and practical purposes. Metallurgy as a science advanced slowly until it was sparked by the invention of the microscope. Although biological microscopy began in the seventeenth century, it was not until two hundred years later that the microscope was employed meaningfully as a tool in metallurgy. Through observations made with the microscope, the crystalline nature of metals was definitively established. In 1863, Henry Sorby, a native of the steel center of Sheffield, England, developed techniques for the study of thin sections of rocks, using polarized light. Fascinated by the crystalline nature of meteorites, he experimented with steels and iron. His microscopic work on metals was not taken seriously until twenty years later when its practical value became apparent. It was Floris Osmond, an engineer working in the last part of the nineteenth and first part of the twentieth century, who combined microscopy and thermochemistry in his work on steel and produced a body of information and procedures that established metallography as a modern science.

Today's metallography employs the optical microscope (as well as the electron microscope and X-ray techniques) to determine the structural properties of metals and alloys. The metallographer studies how constituent crystals act when a given metal is subjected to different stresses, to fracture, and to variations in temperature. The specimen is always observed carefully during the process of prolonged heating and slow cooling. This data is plotted on a chart (equilibrium diagram), which is used to provide information about melting points, solidification, changes in physical structure, and so on. Among other things, through such metallographic techniques it is possible to prepare alloys that will meet special requirements. For example, it is essential for some metals to retain their strength and form when exposed to great stresses, as in the case of turbine blades in jet engines or guided missiles. Metallography helps to determine the kind of metals and metal alloys required to withstand such stresses.

In order to examine a metal specimen properly with the microscope, a clean and fresh area of the surface must be exposed. This process is laborious and consists essentially of subjecting the metal to emery papers of decreasing coarseness until the fine scratches can be removed by polishing to a mirror-like finish with alumina or with chromic acid. In some instances electrolytic polishing is preferable. The polished specimen is then immersed in any of several corrosive chemical solutions that attack certain parts more readily than others and thus create minute hills and valleys on the surface of the metal.

The specimen is then examined through an optical microscope lit by vertical illumination, or, in large metallurgical laboratories, by a specially designed metallographic microscope that permits rapid viewing. The surface depressions and elevations caused by the etching process reflect different intensities of light or, with polarization, different color values. The specimen is magnified anywhere from one hundred to fifteen hundred diameters. Where greater magnifications are required, an electronic microscope is used.

Pure minerals

Minerals are best described as inorganic substances occurring naturally in the earth, each having a set of unique physical properties, classified according to color, hardness, and crystalline organization. The earth's crust is a prodigious source of minerals that have served man's practical and ornamental needs since the beginnings of civilization. More than fifteen hundred different rock species have been identified, each amply described and classified in the expanding geological science of minerology. Minerals and derivatives of minerals are of great economic importance to man. The major categories of minerals, with a few examples of each, are gems (diamond, garnet, opal), structural materials (calcite, gypsum), ceramics (feldspar, quartz), fertilizers (apatite, sylvite), pigments (hematite, limonite), and metalliferous ores (molybdenite, chromite, cinnabar).

The microscope has revealed the inner architecture of minerals and has helped to delineate their crystalline structure. The microscope has thus paved the way for the new science of *structural crystallography* by providing a basis for classifying the internal symmetries and dimensions of minerals.

The photomicrographs in this chapter illustrate some typical mineral samples. They are made from extremely thin sections of specimens and illuminated by transmitted polarized light, which activates the crystalline content and enhances the beautiful shapes and variegated colors.

Attempts at classification of minerals go back as far as the fourth century B.C., when the Greek philosopher Theophrastus, a pupil of Aristotle, wrote a basic treatise on the nature of minerals. The most exhaustive ancient discourse on the mineral kingdom, however, appeared in the last five books of *Natural History* by Pliny, about A.D. 77. Understandably, these studies were largely descriptive, rather than analytical, since scientific research was still in its infancy. It was not until 1546 that Georgius Agricola in his *De natura fossilium* classified minerals systematically on the basis of their physical properties. His work was later advanced through the development of chemical methods of analysis, initiated about 1800 by M. H. Kalproth and aided by R. J. Haiiy's studies of crystallography. The polarizing and electron microscopes, and the new X-ray and electron diffraction techniques—along with further advances in crystal chemistry—have given us more precise information regarding the nature and properties of the mineral species.

The three main methods of mineral classification are by physical properties (for example, hardness, luster, color, specific gravity, and refraction); by quality of crystalline formation; and by chemical composition (for example, pure elements, sulfides, oxides, halides, phosphates, silicates, sulfates, carbonates, sulfo salts).

Minerals, like plants and animals, are products of the evolution of the universe. Thus the classification and analysis of mineral substances found on other heavenly bodies—such as the moon—can in turn teach man a great deal about the evolution of the earth. Recent advances in space exploration are giving scientists new opportunities to examine and test many "new" substances previously unavailable on our planet.

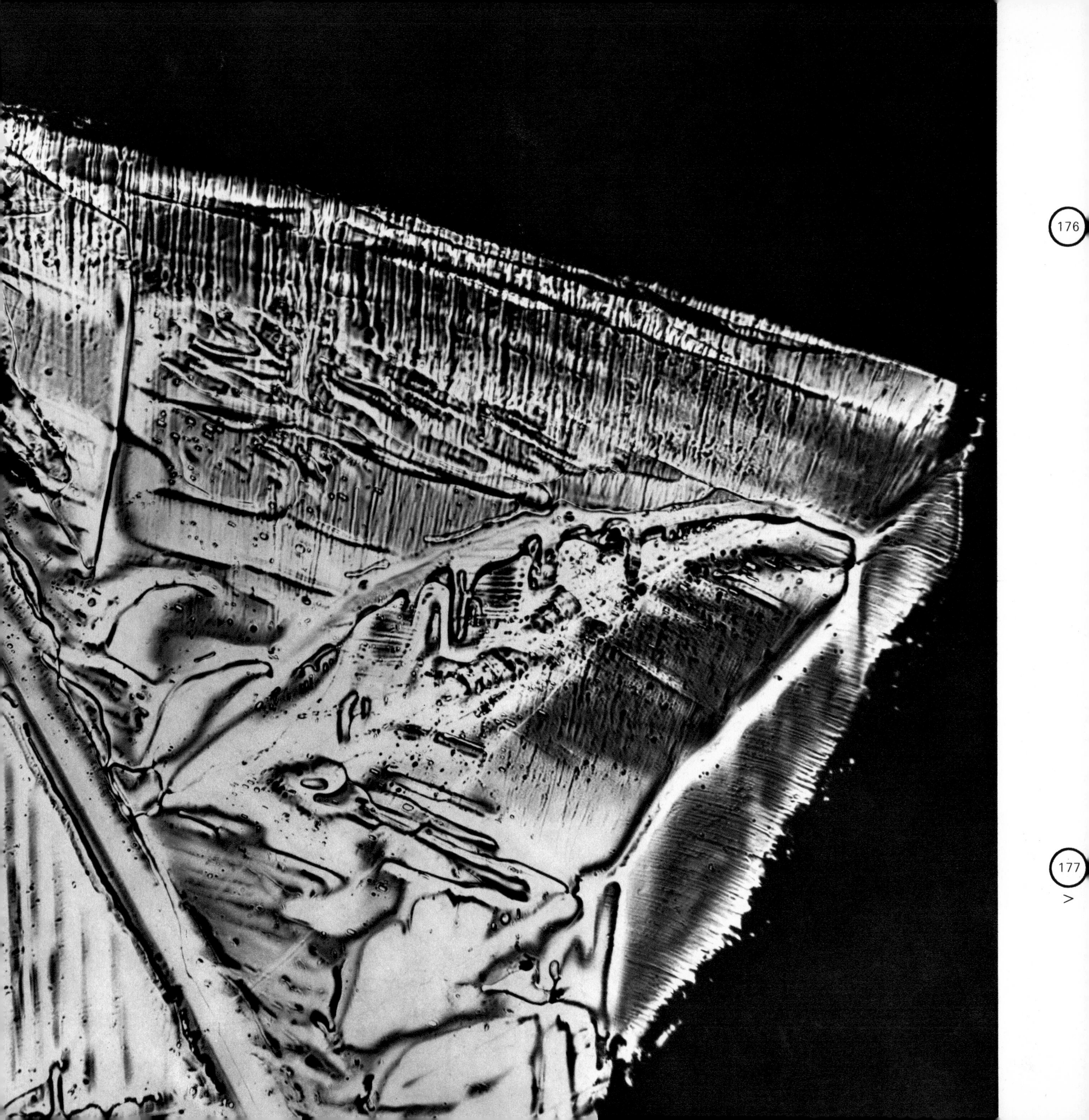
176
177
>

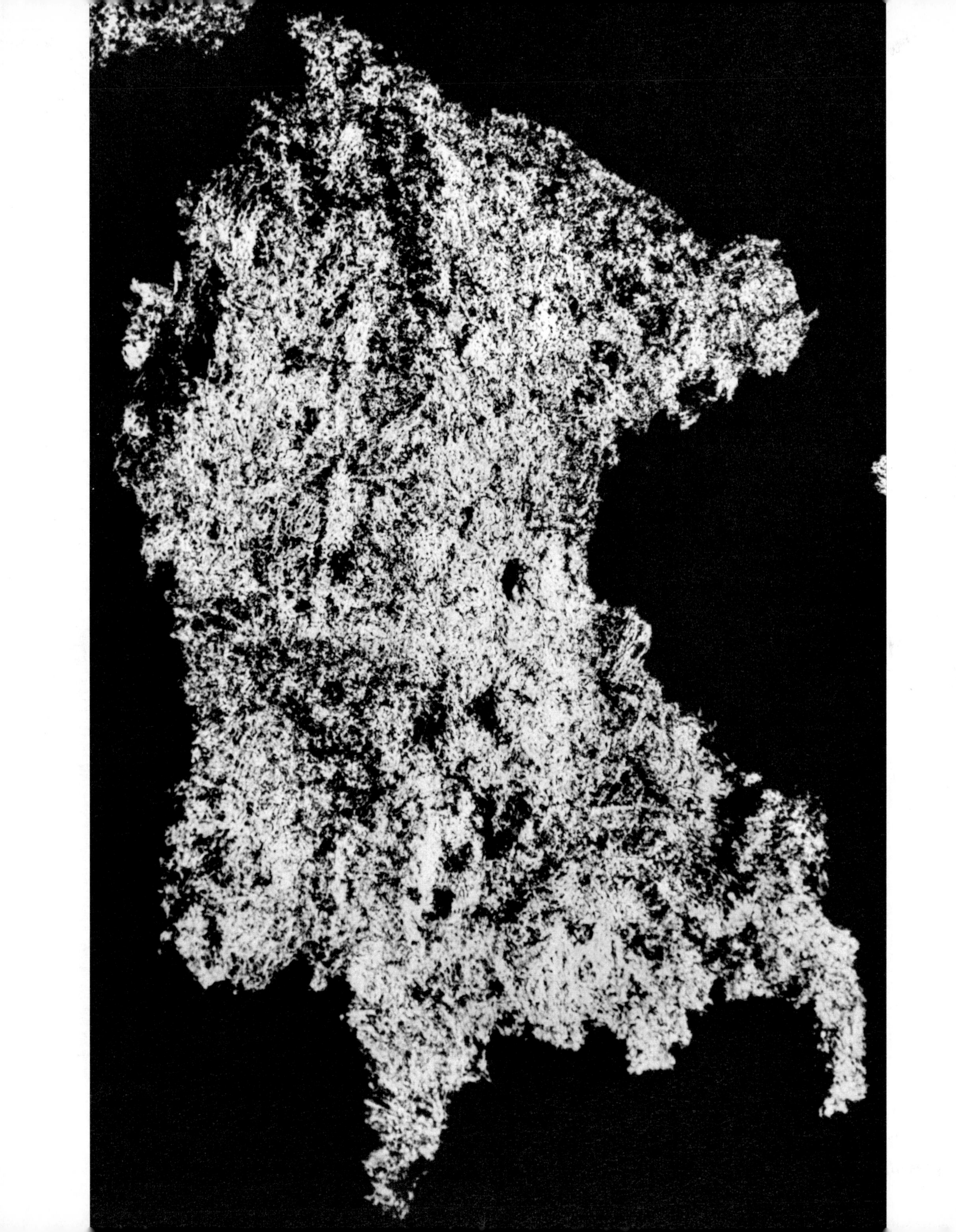

180

181

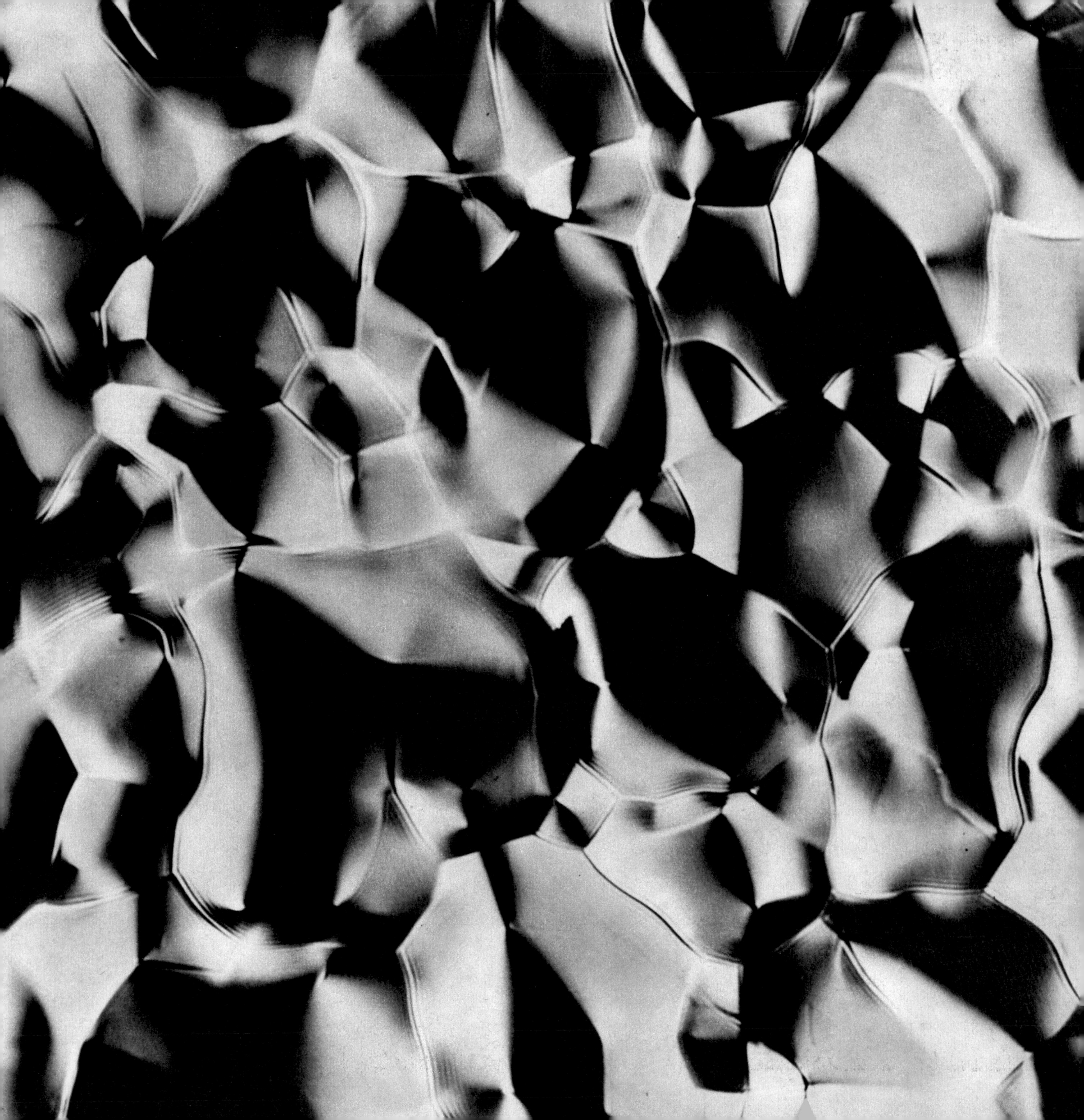

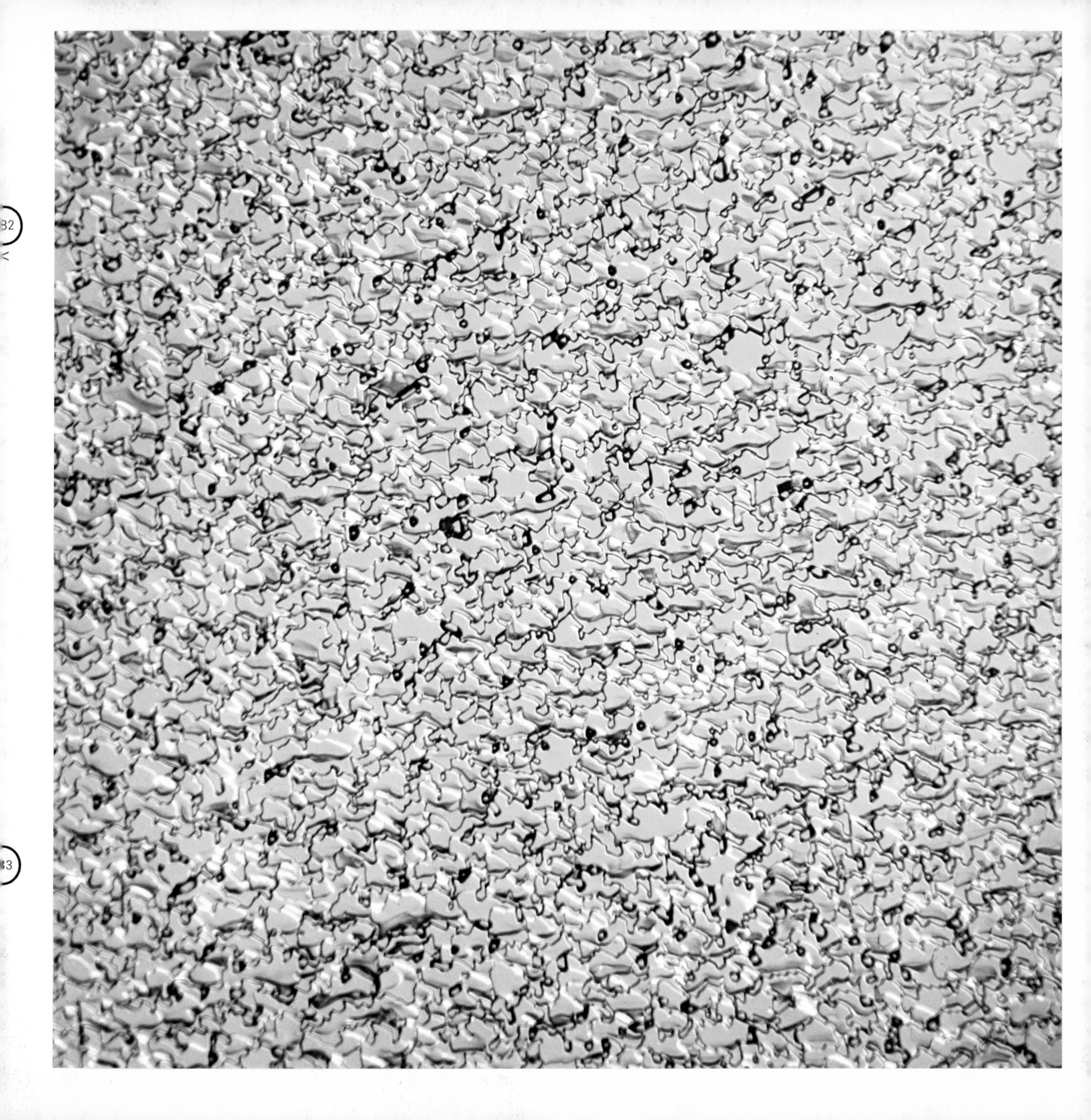

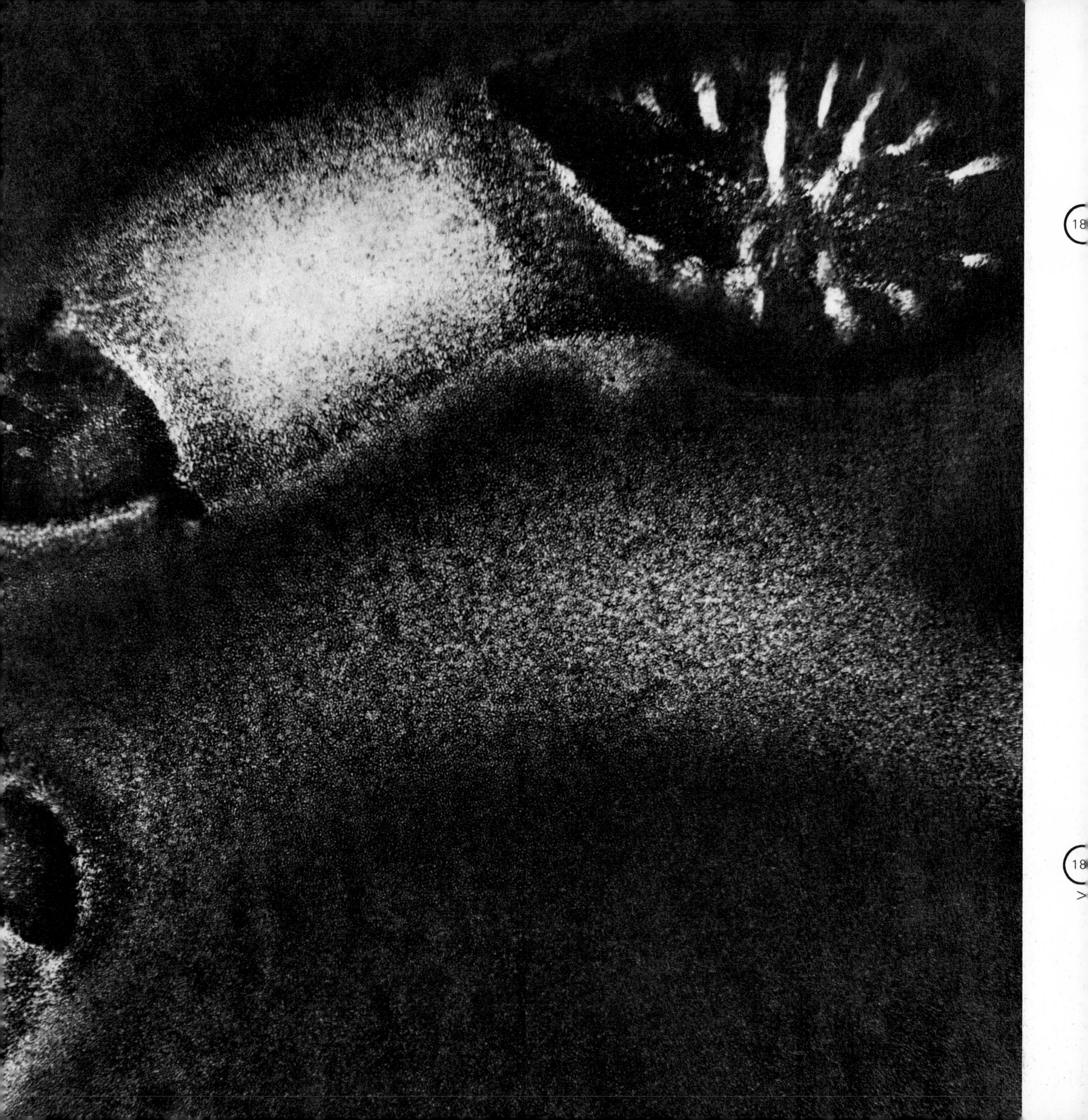

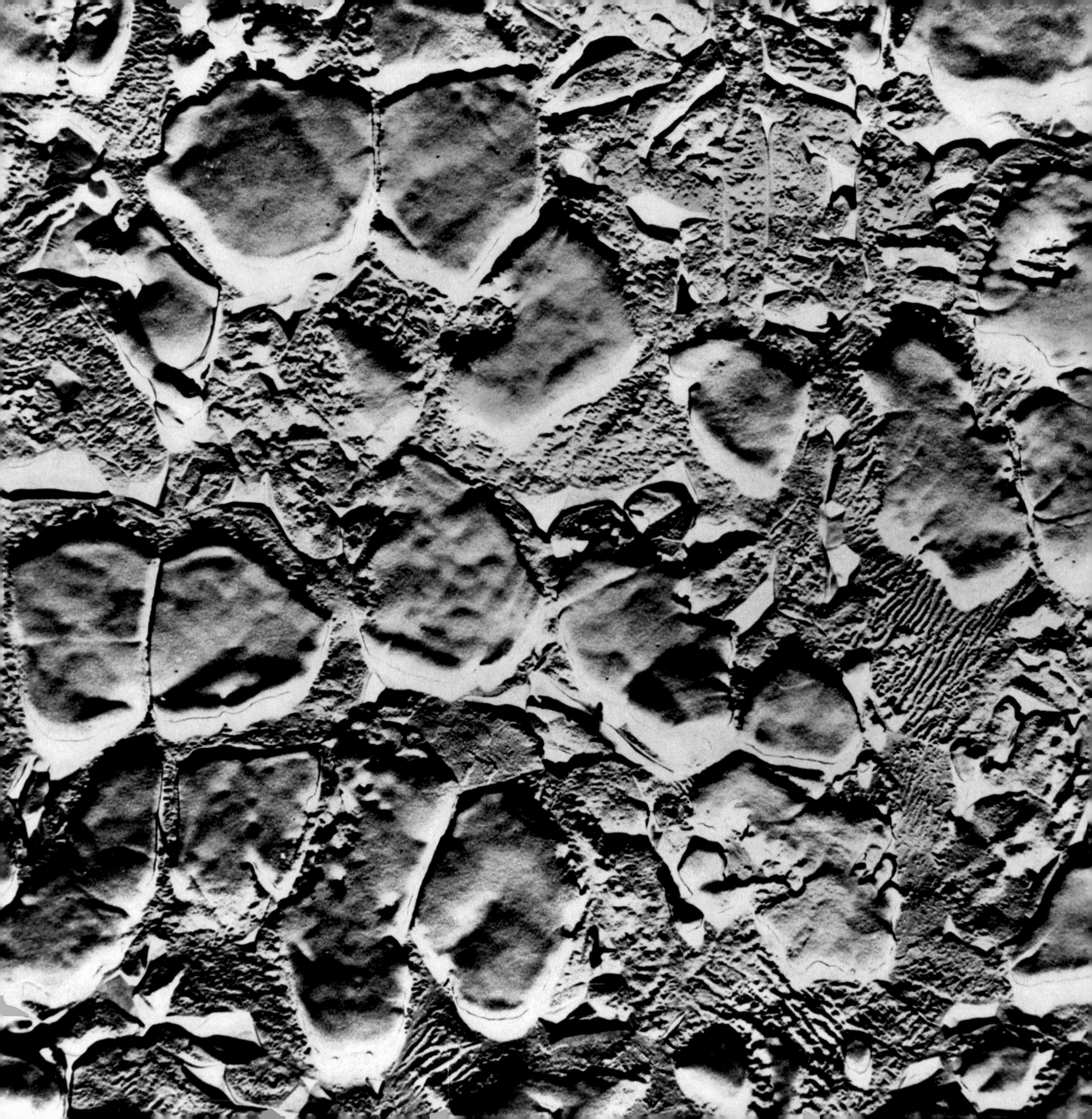

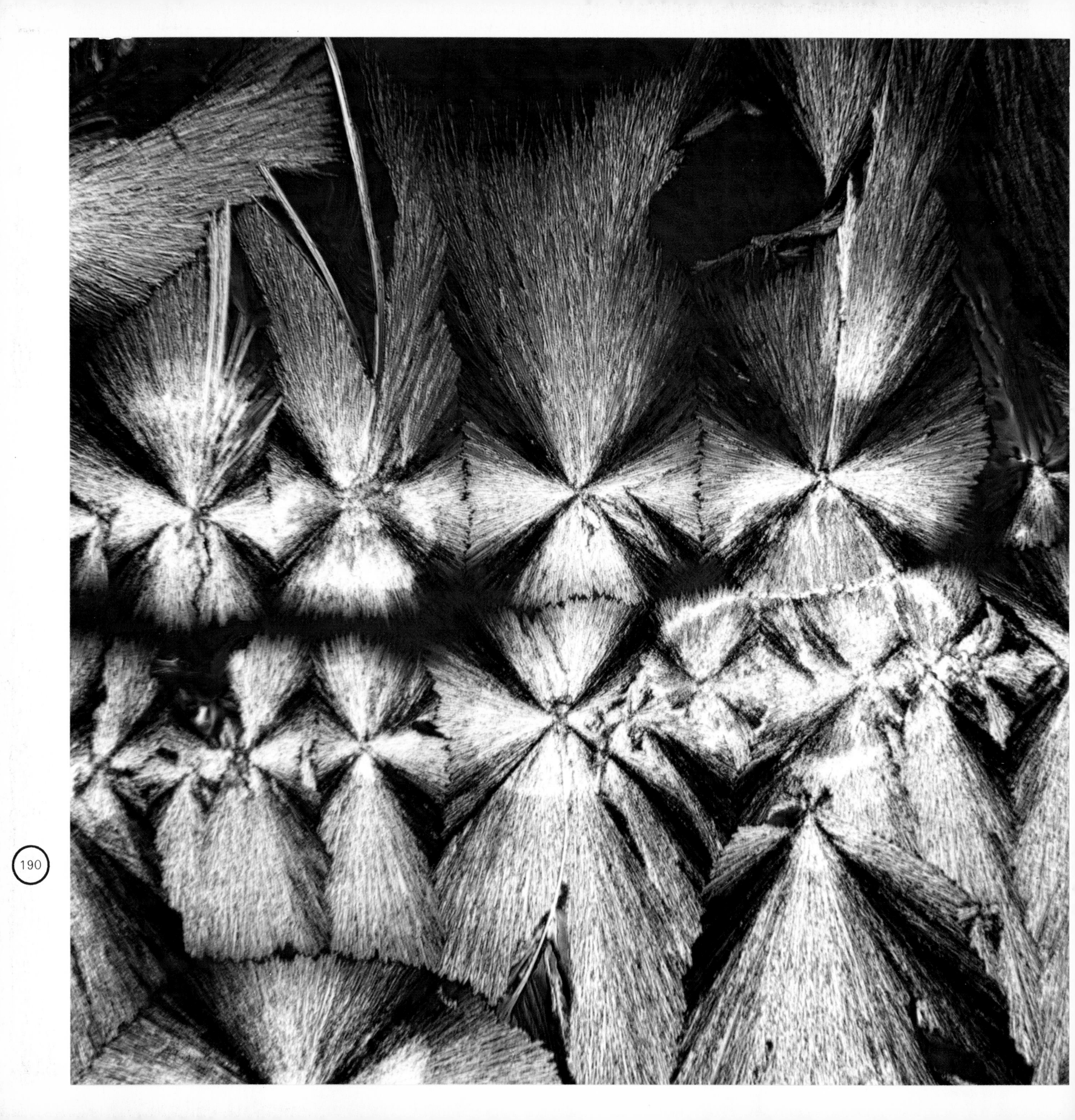

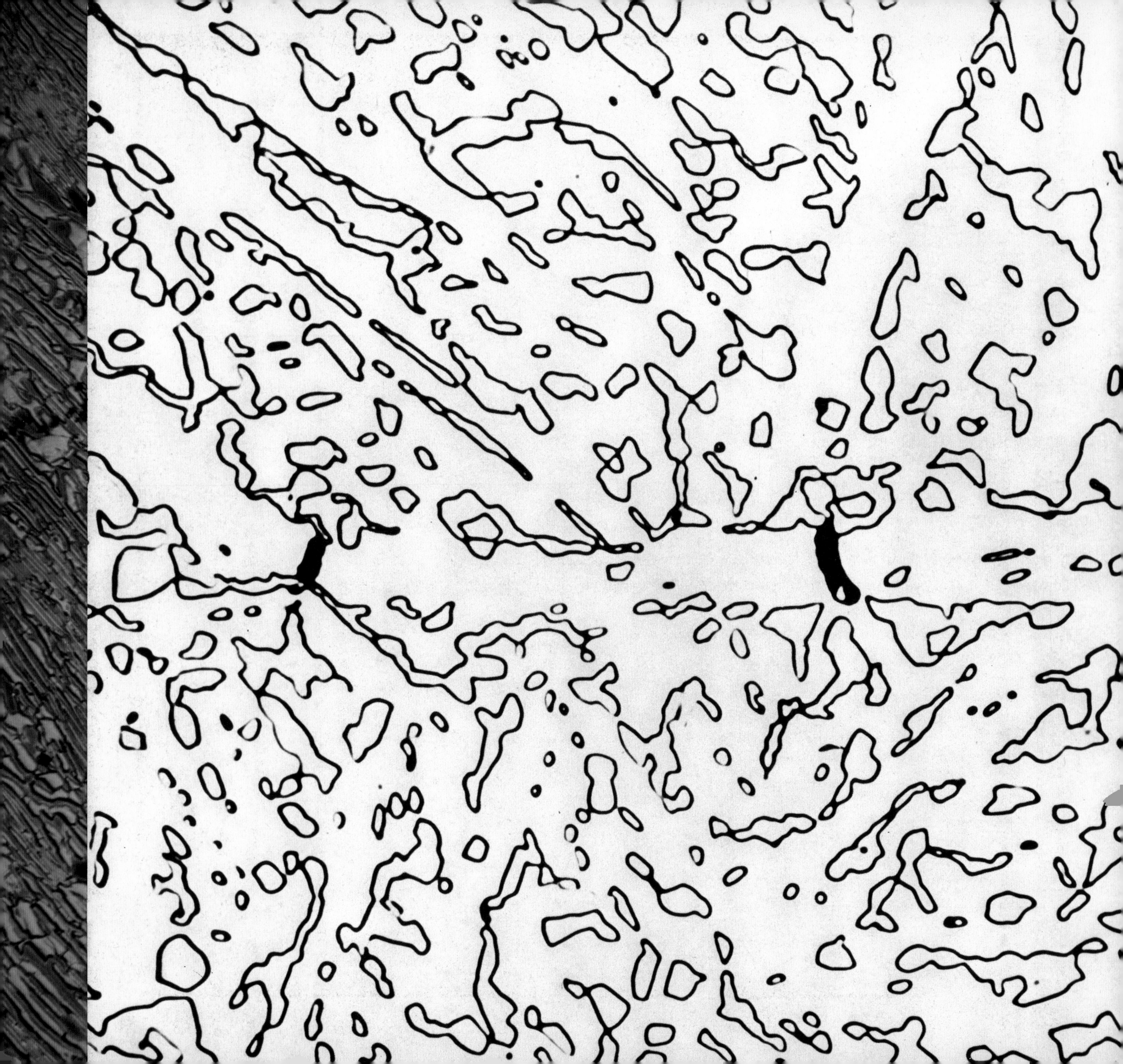

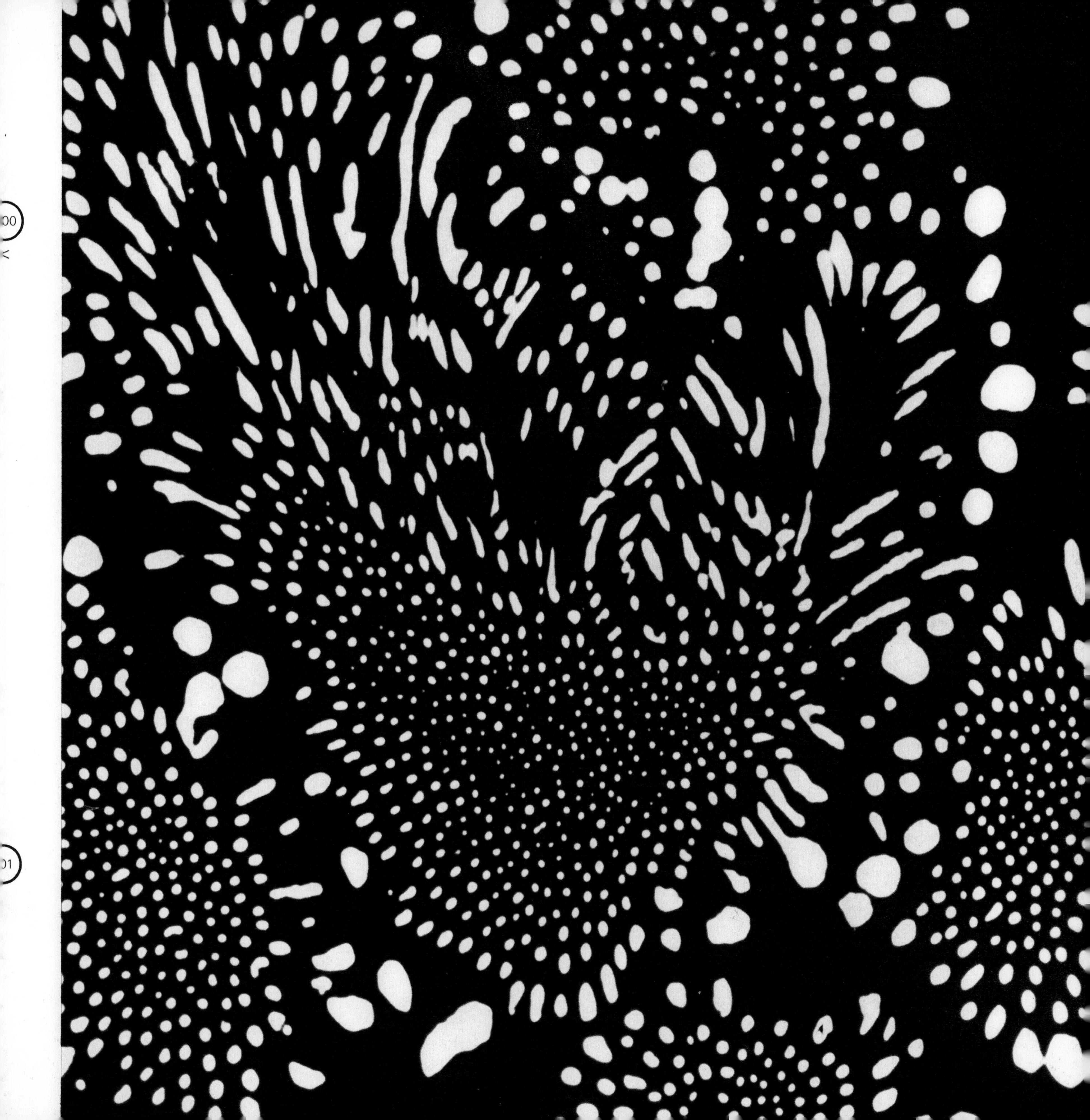

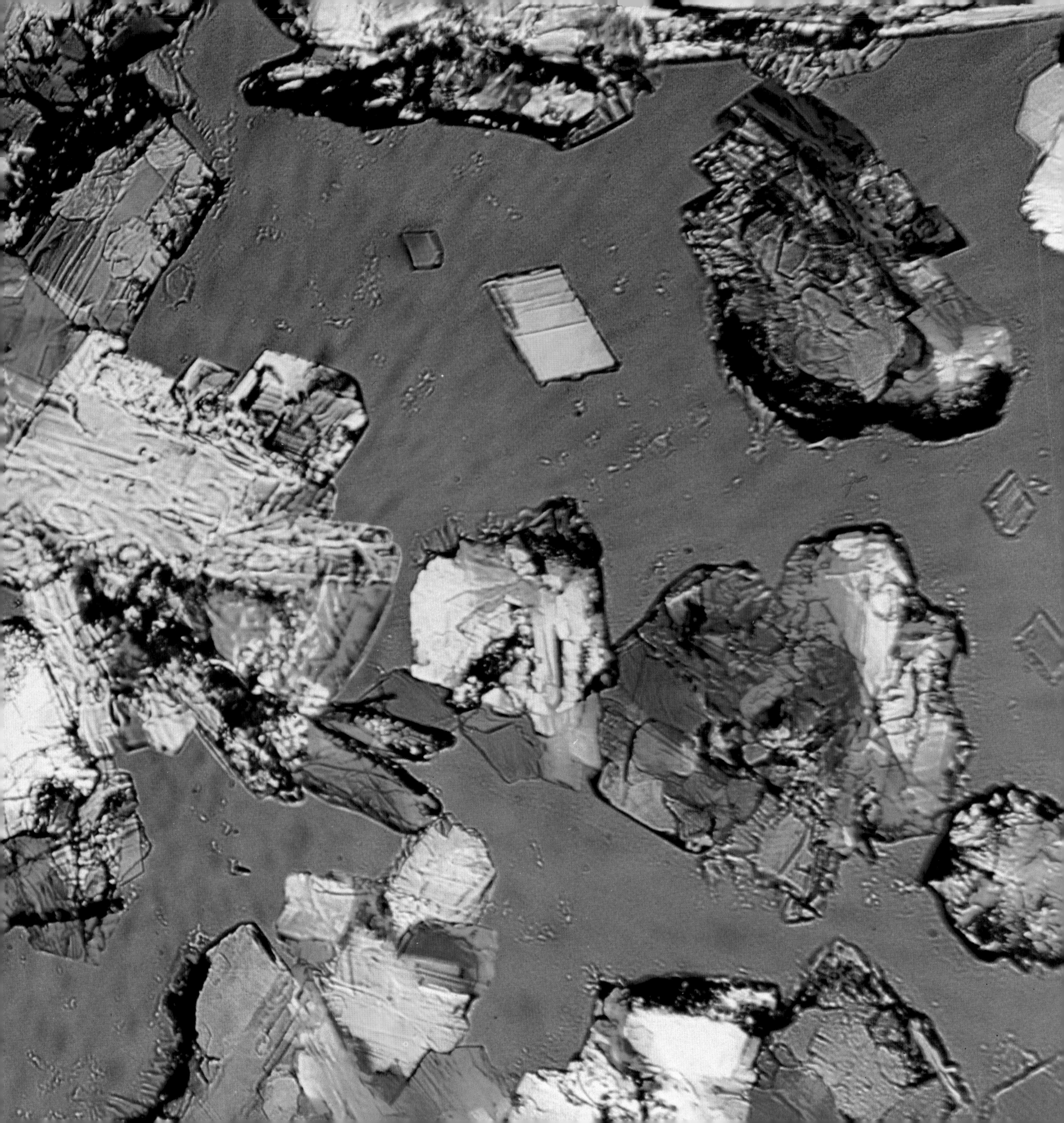

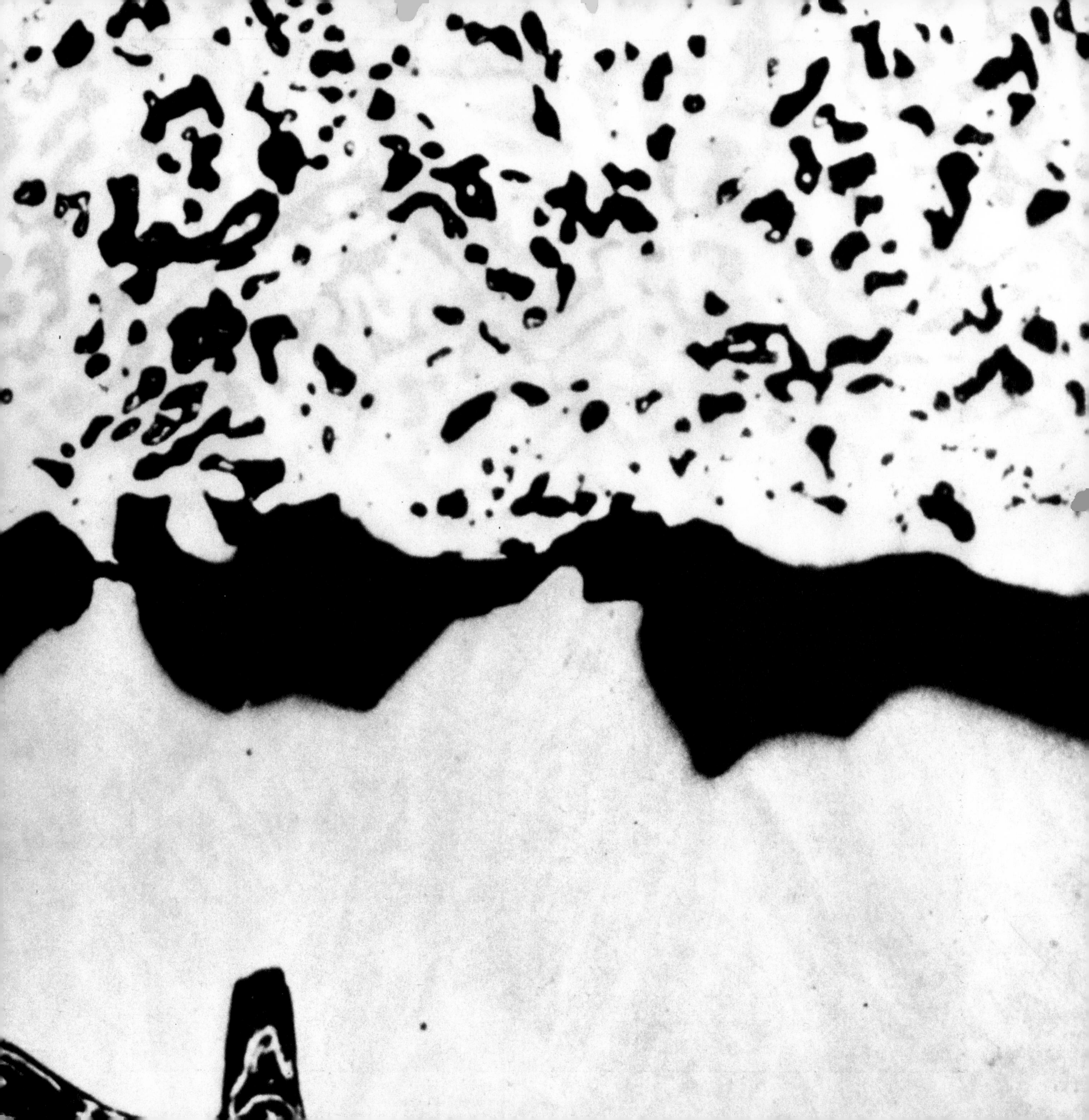

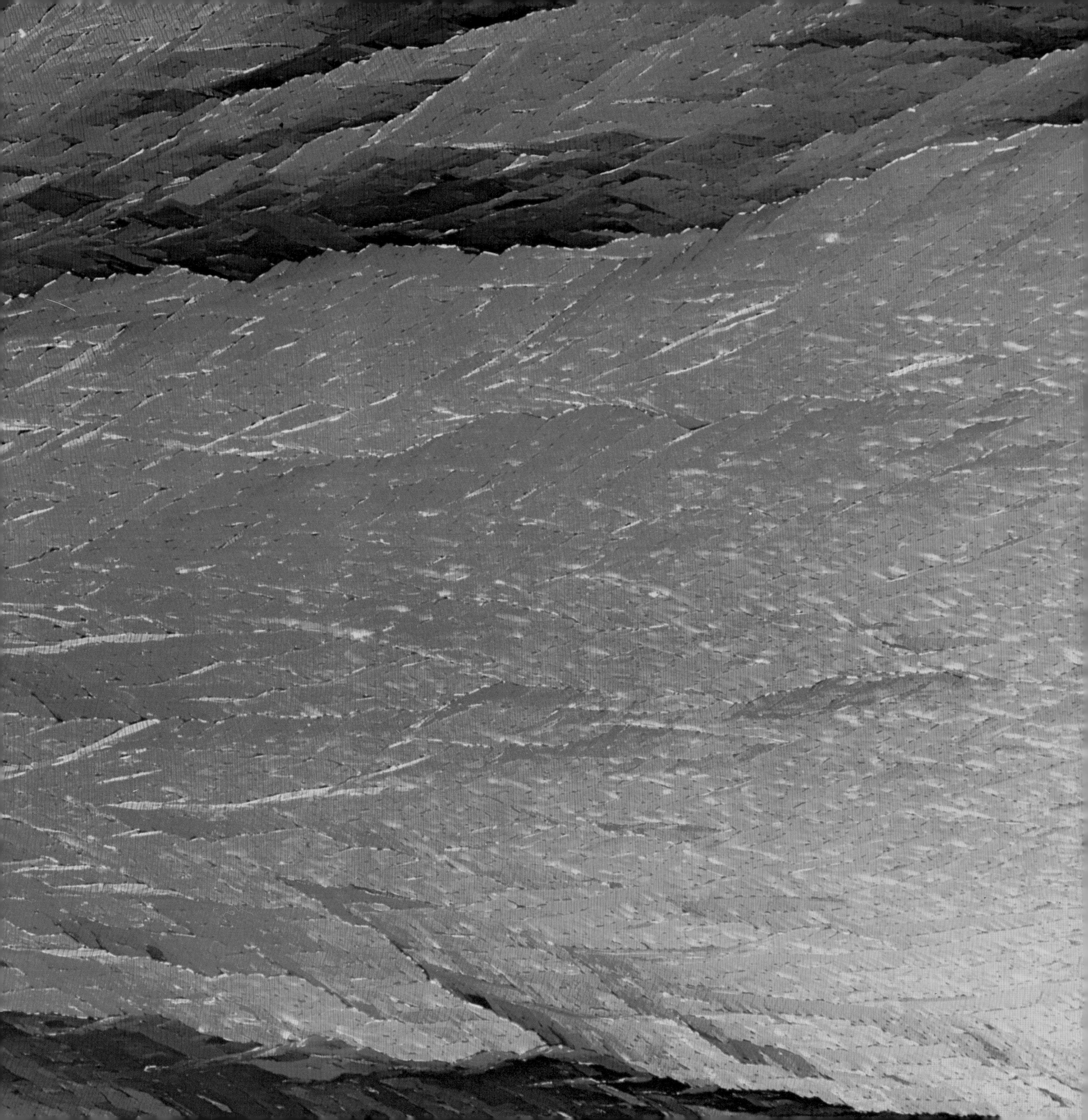

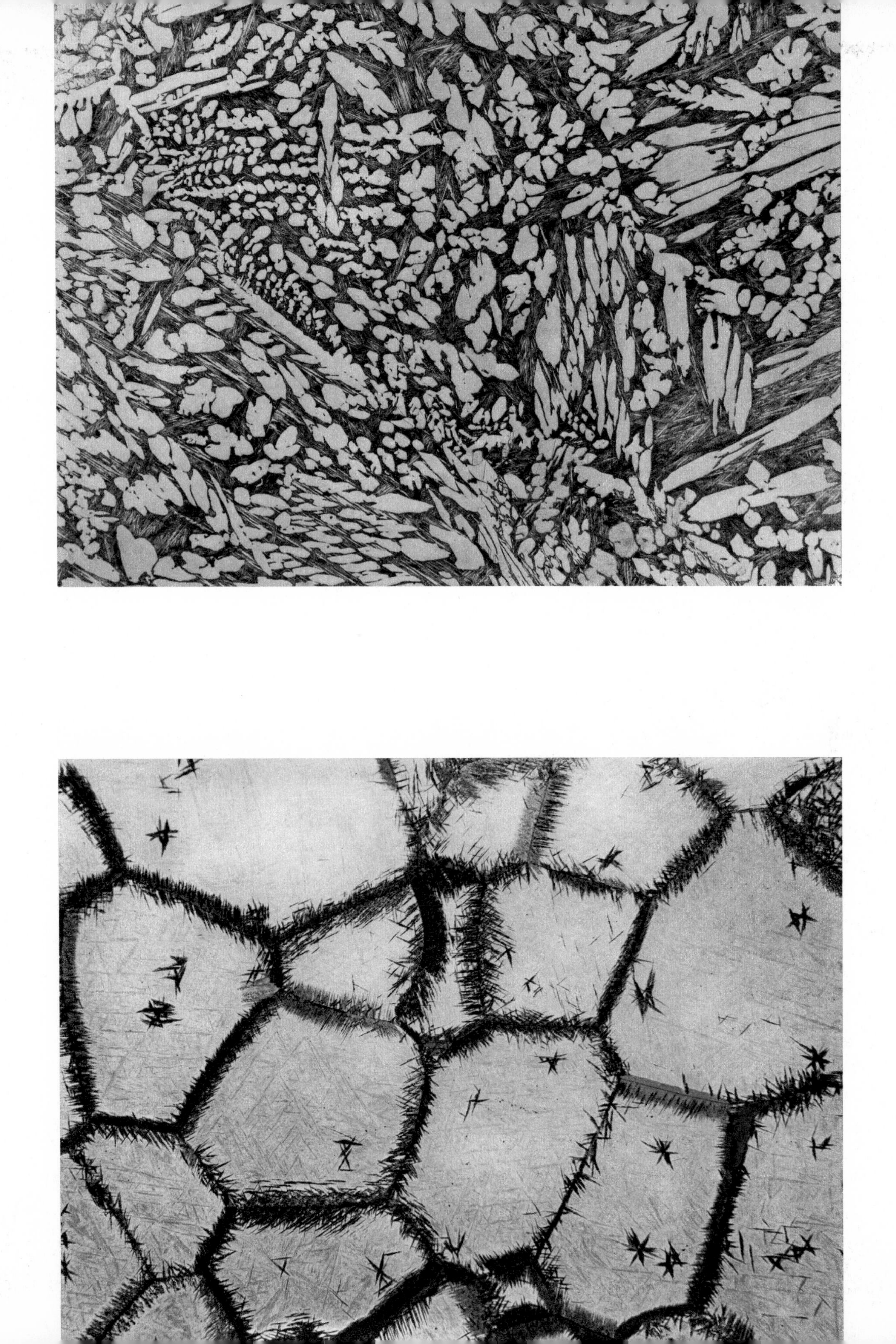

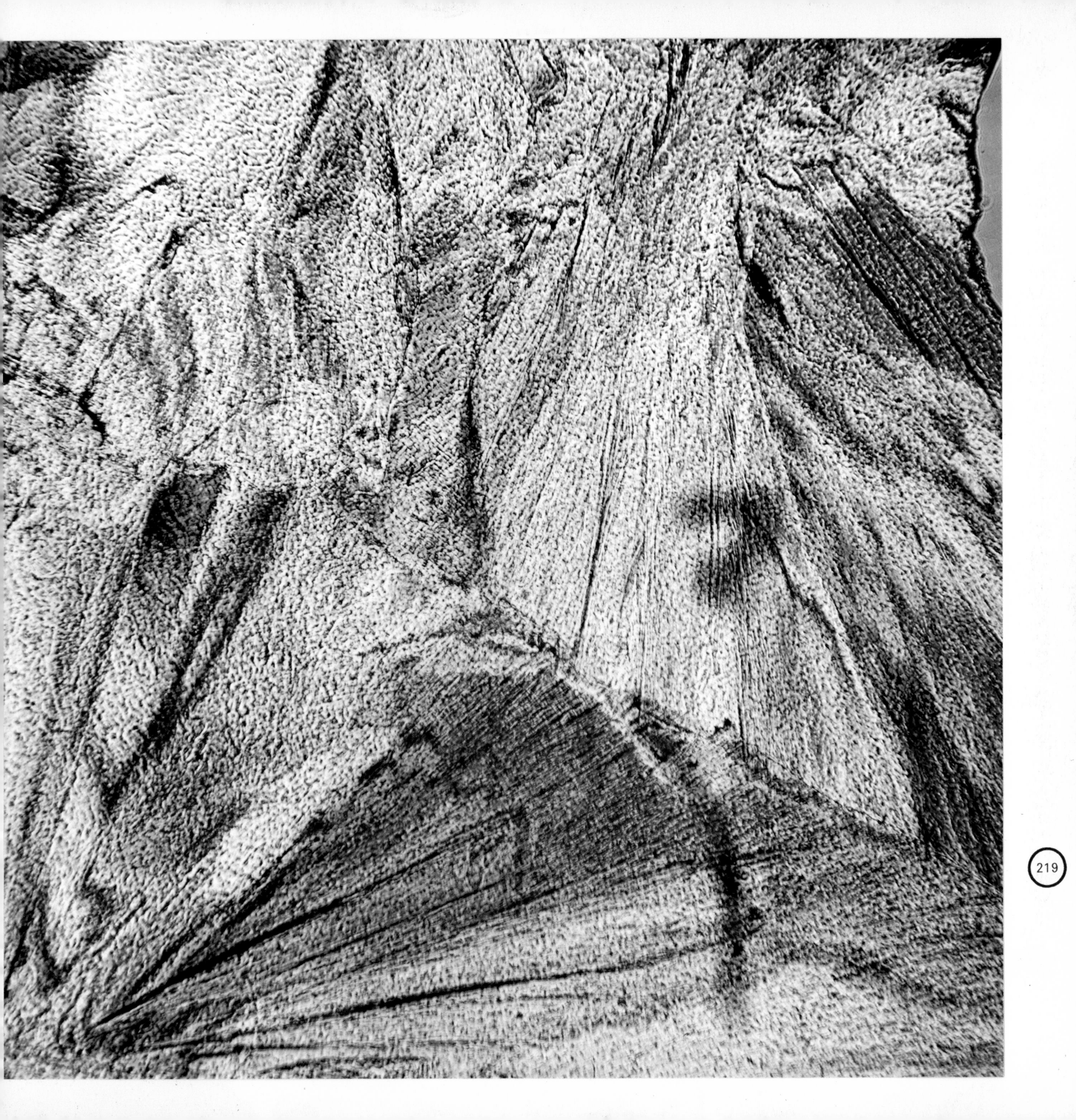

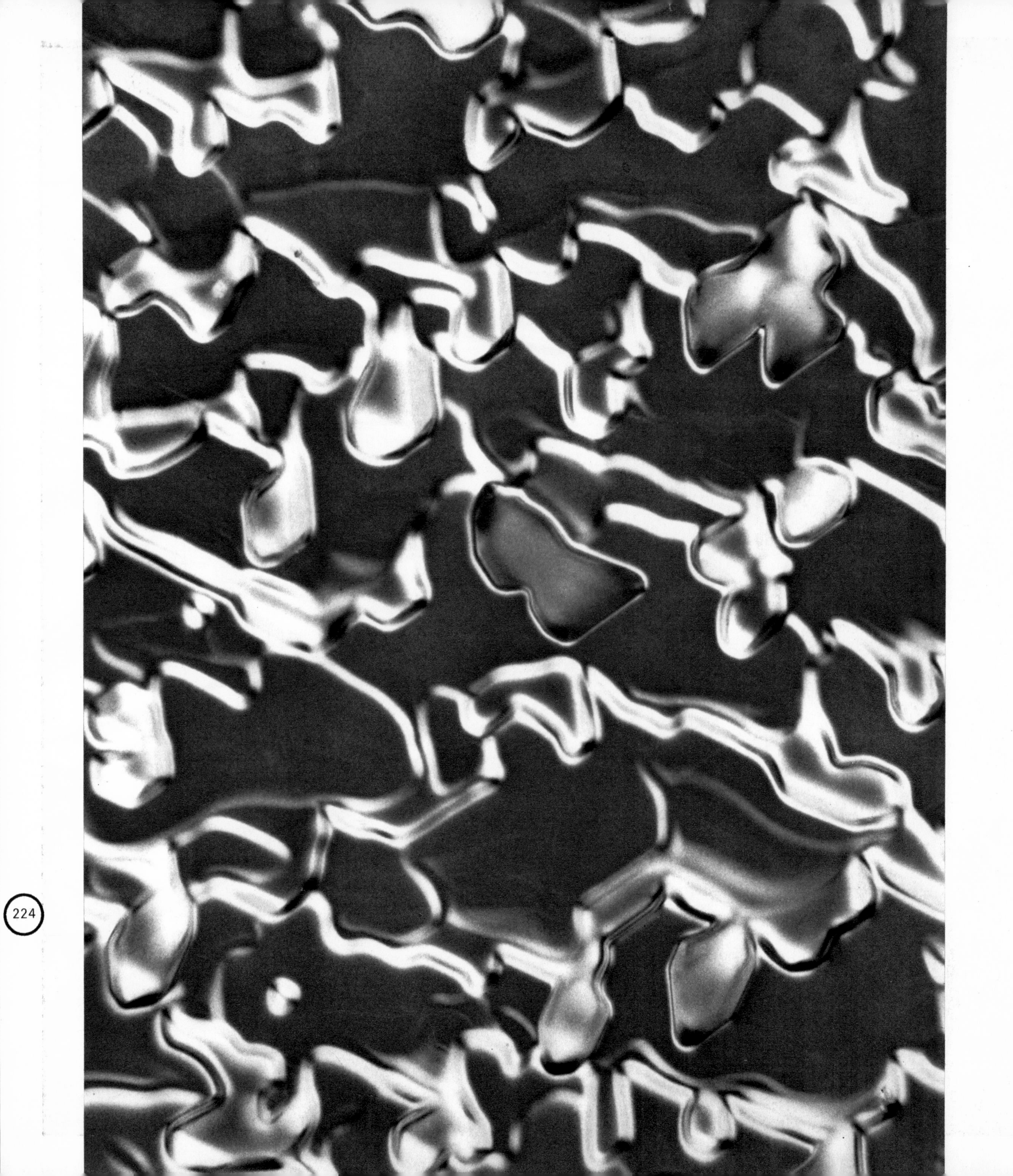

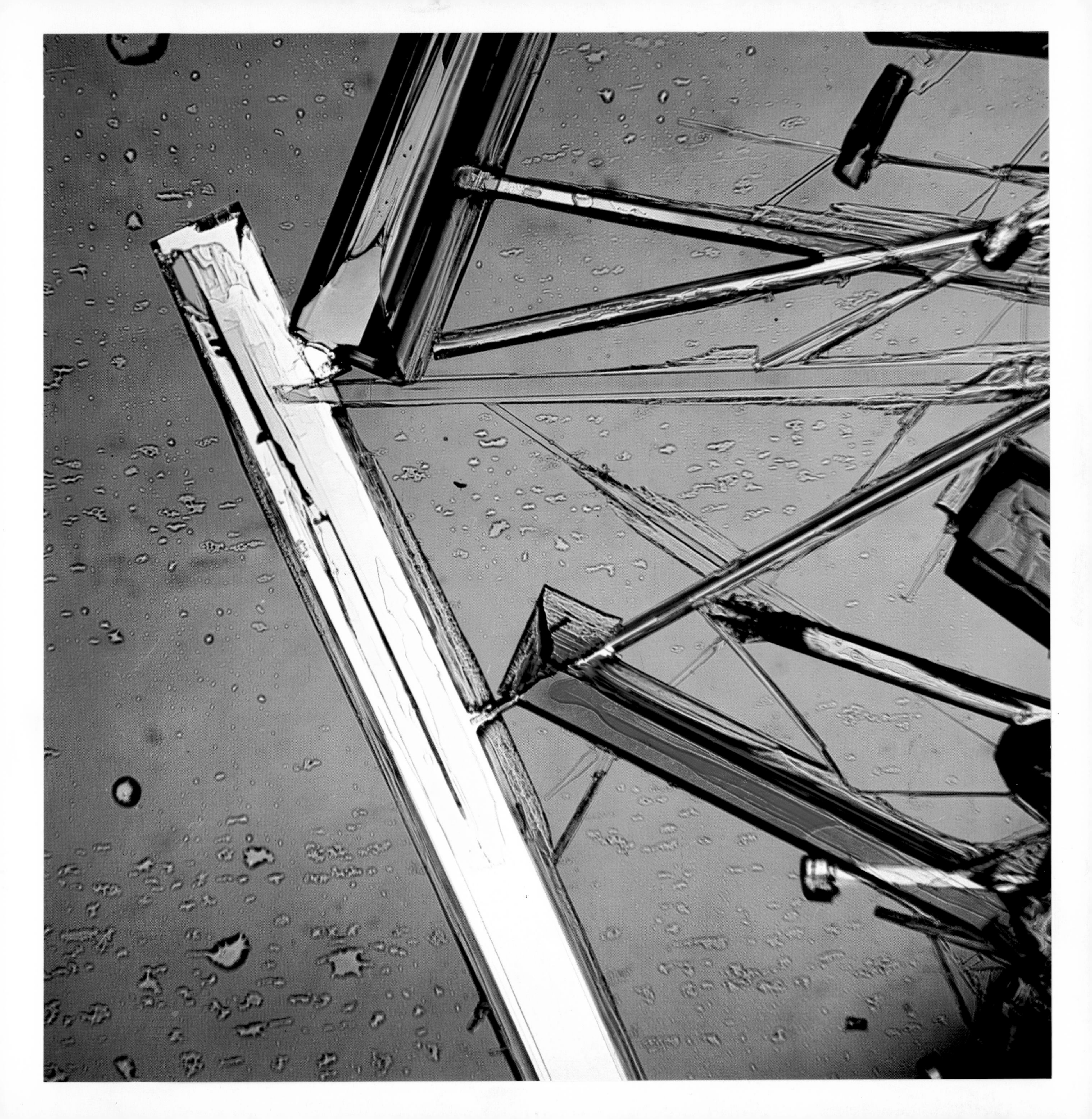

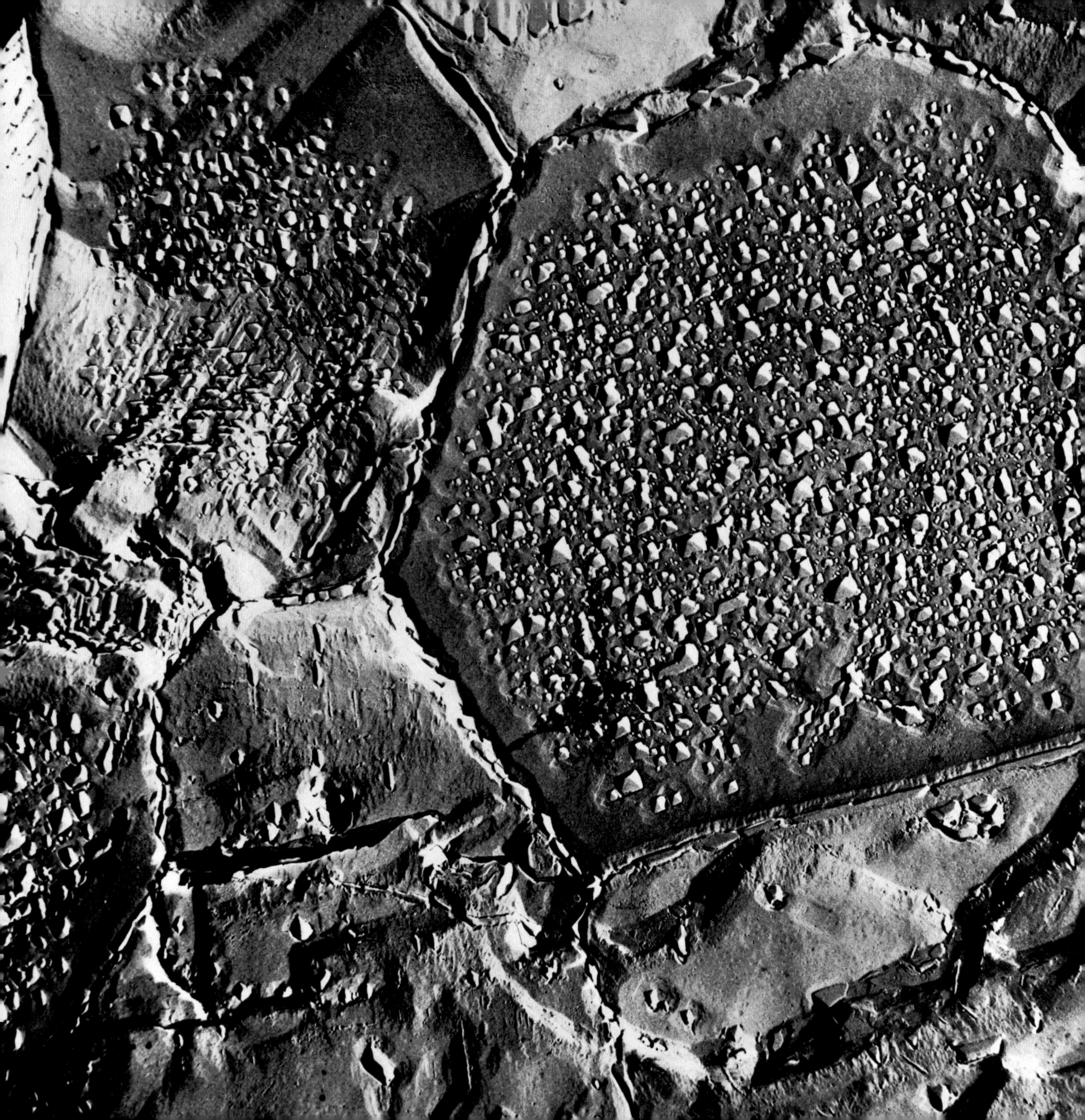

228

229

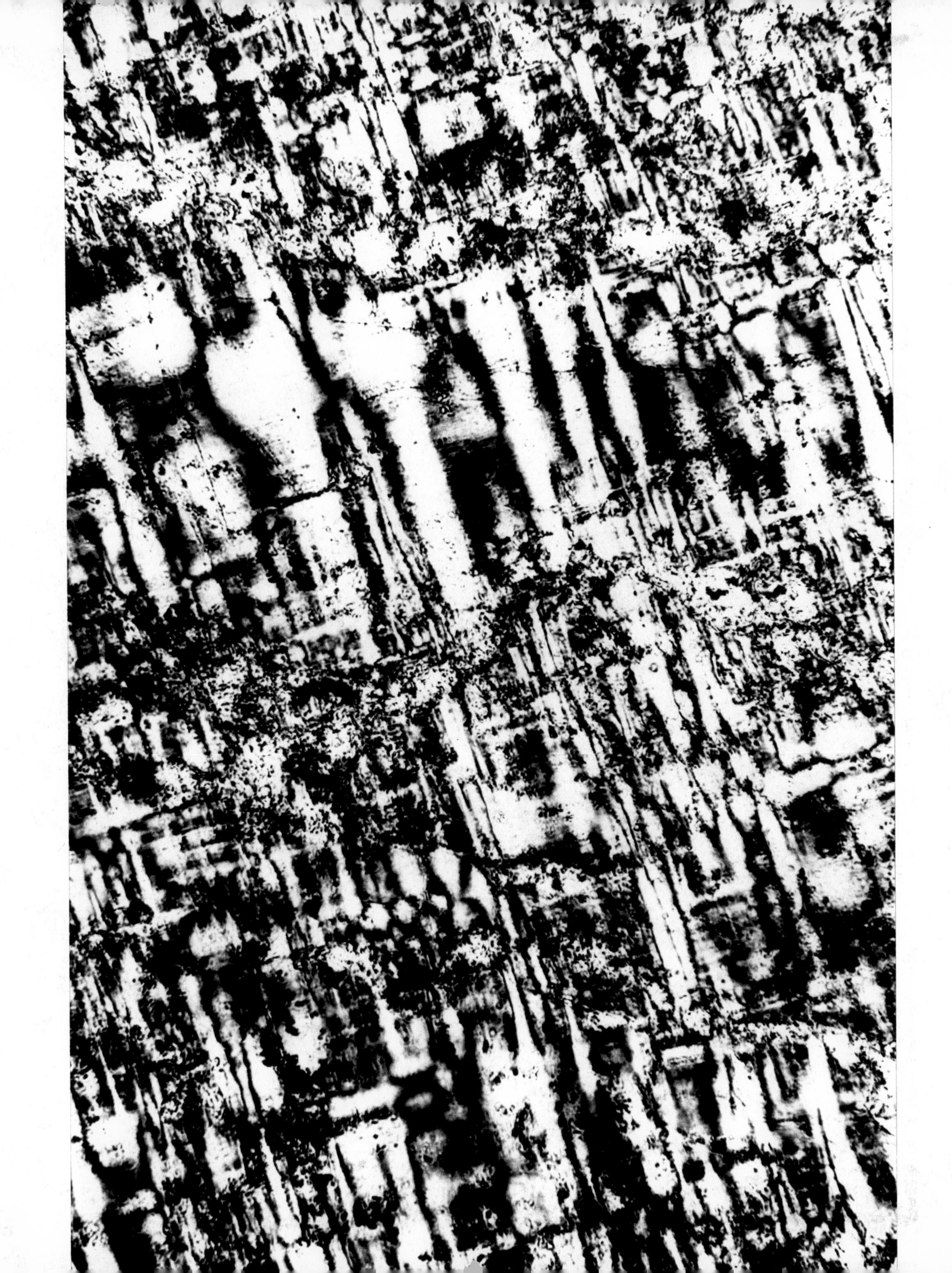

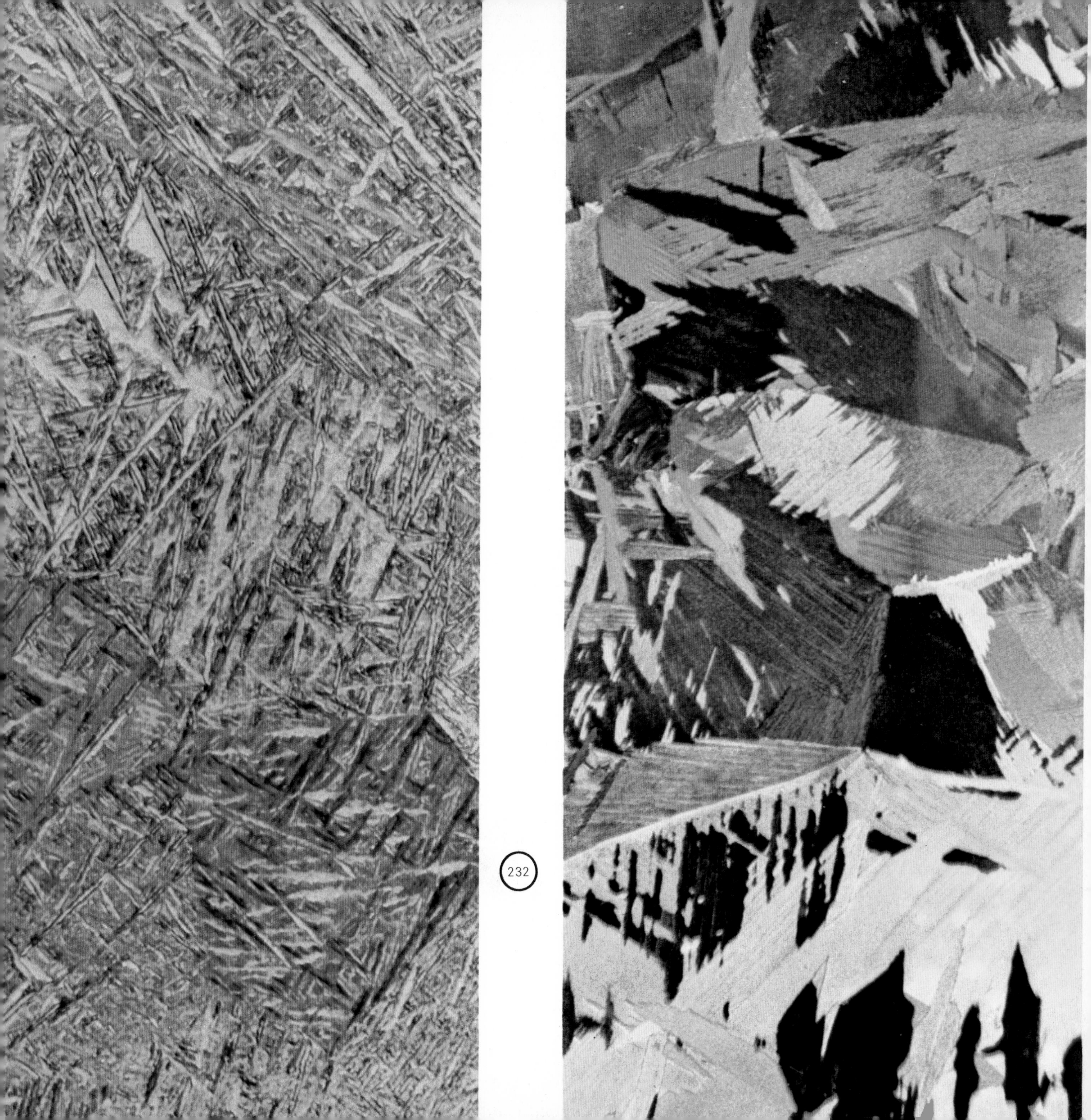

232

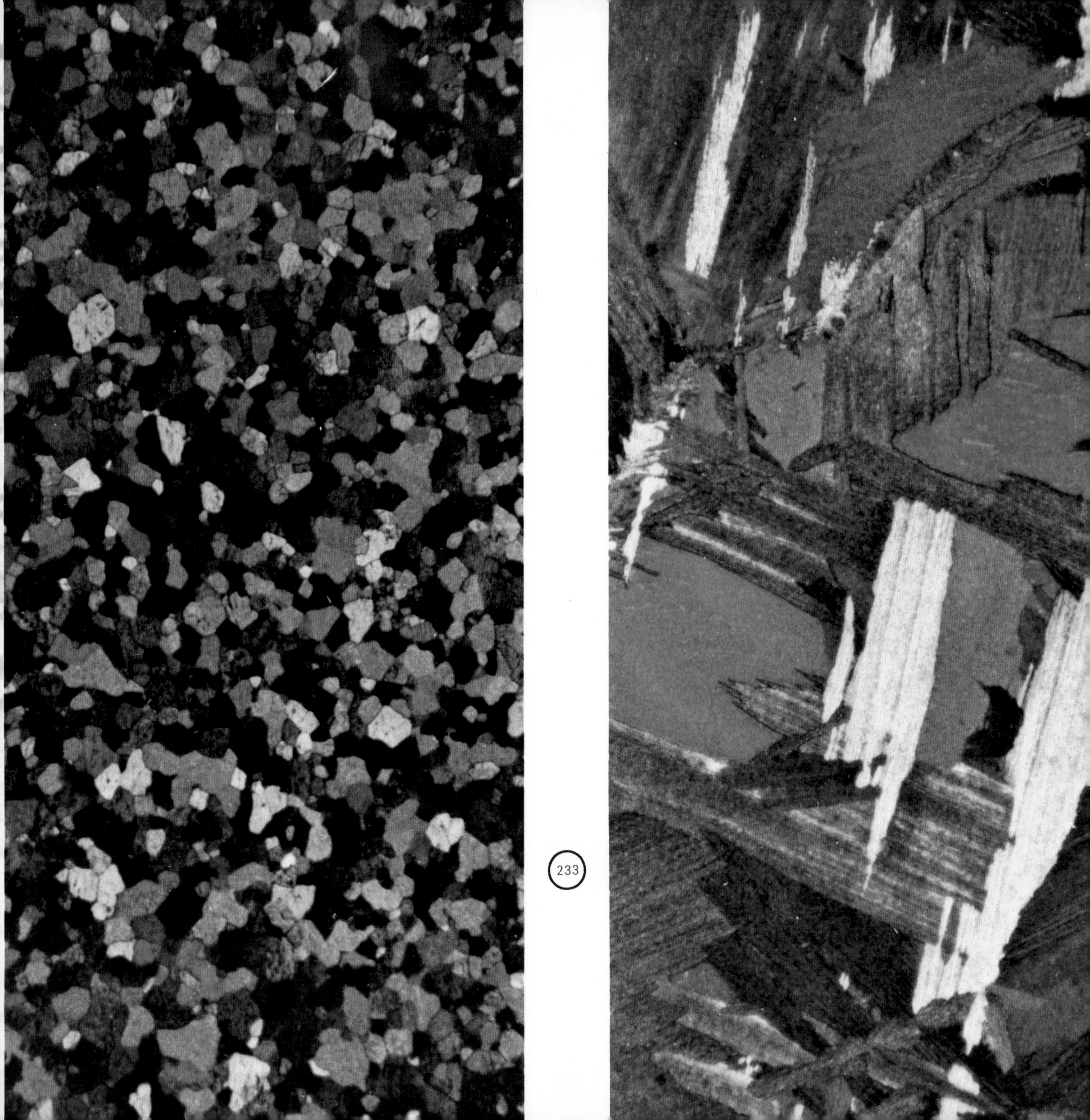

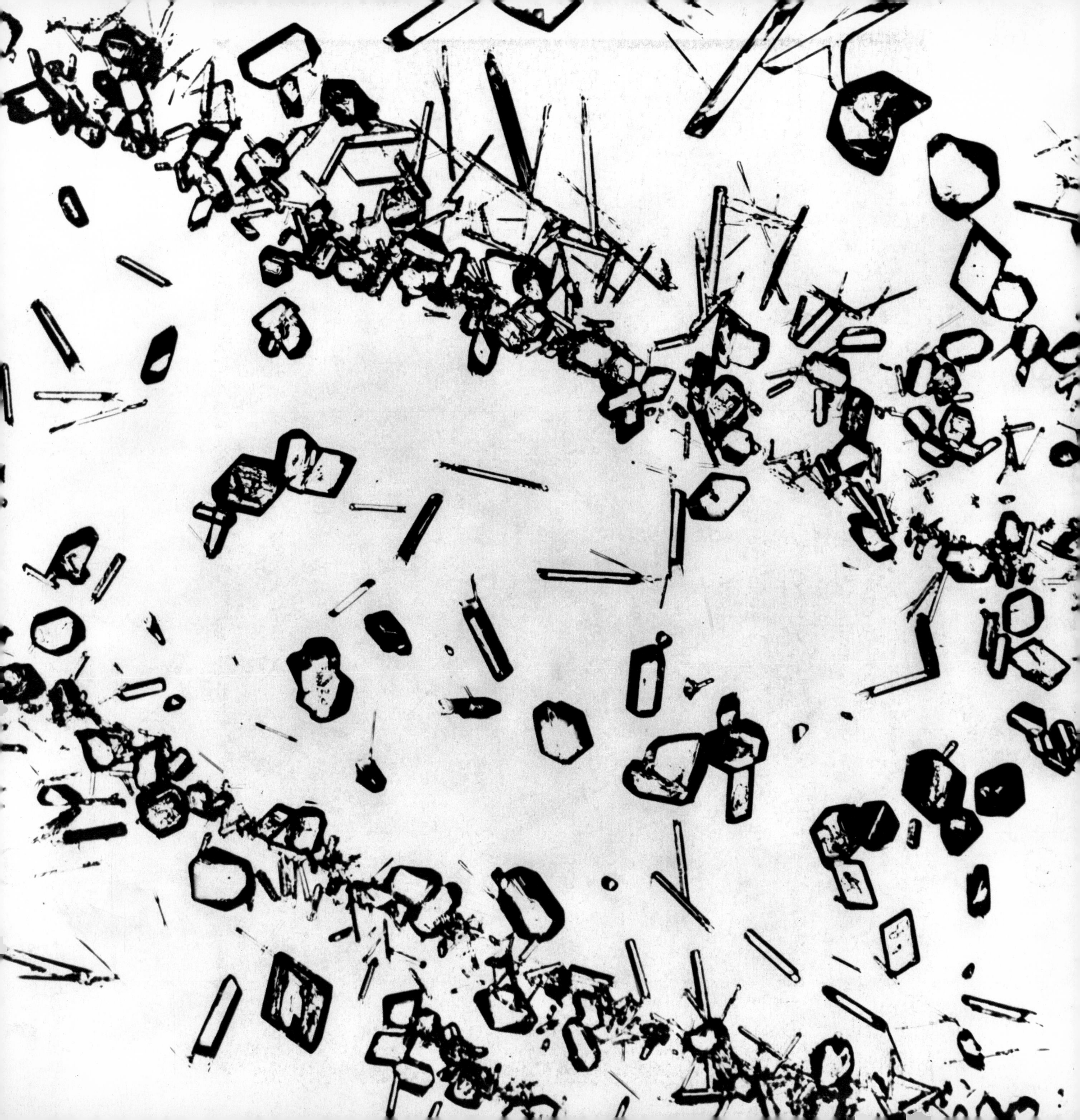

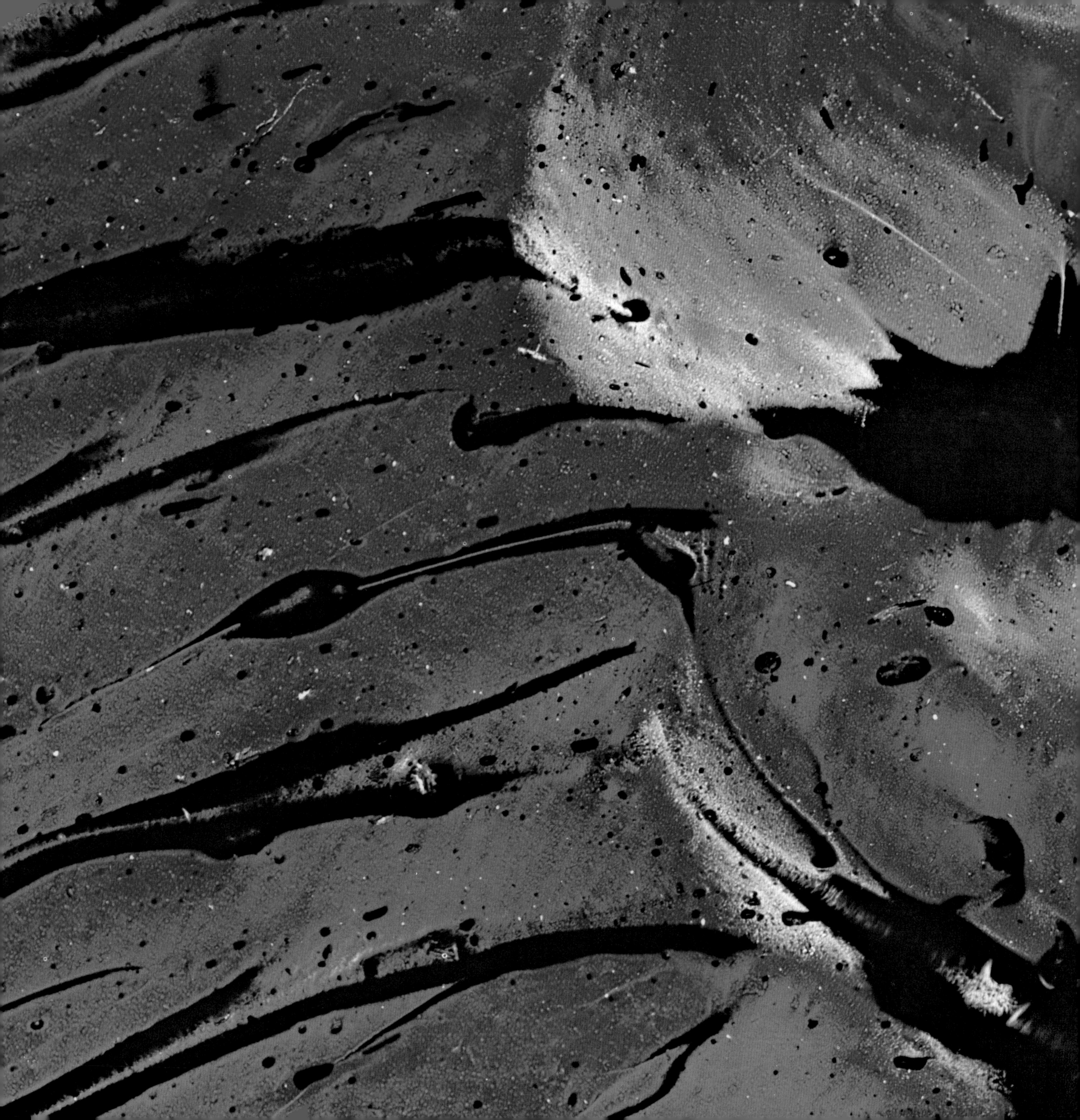

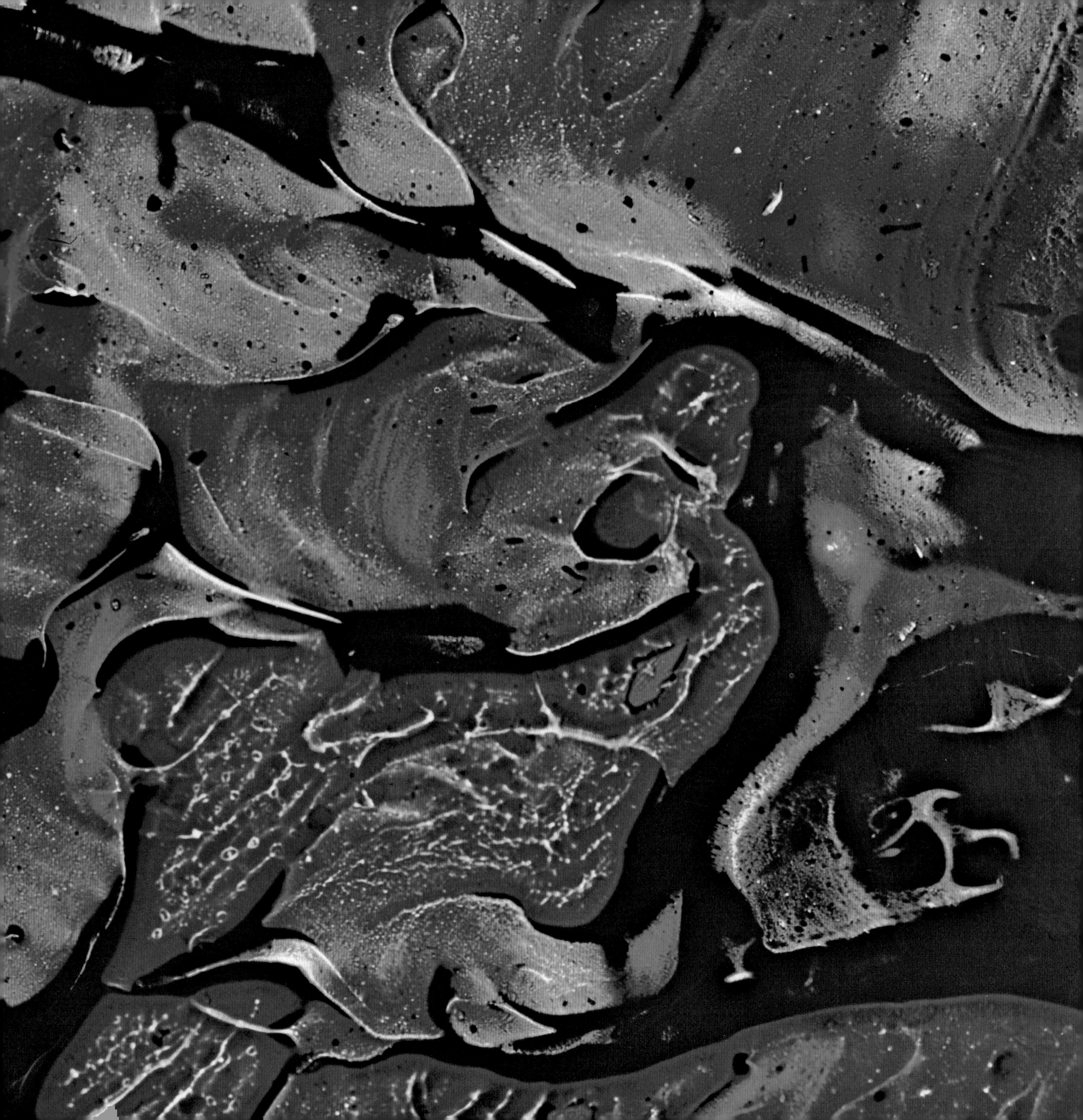

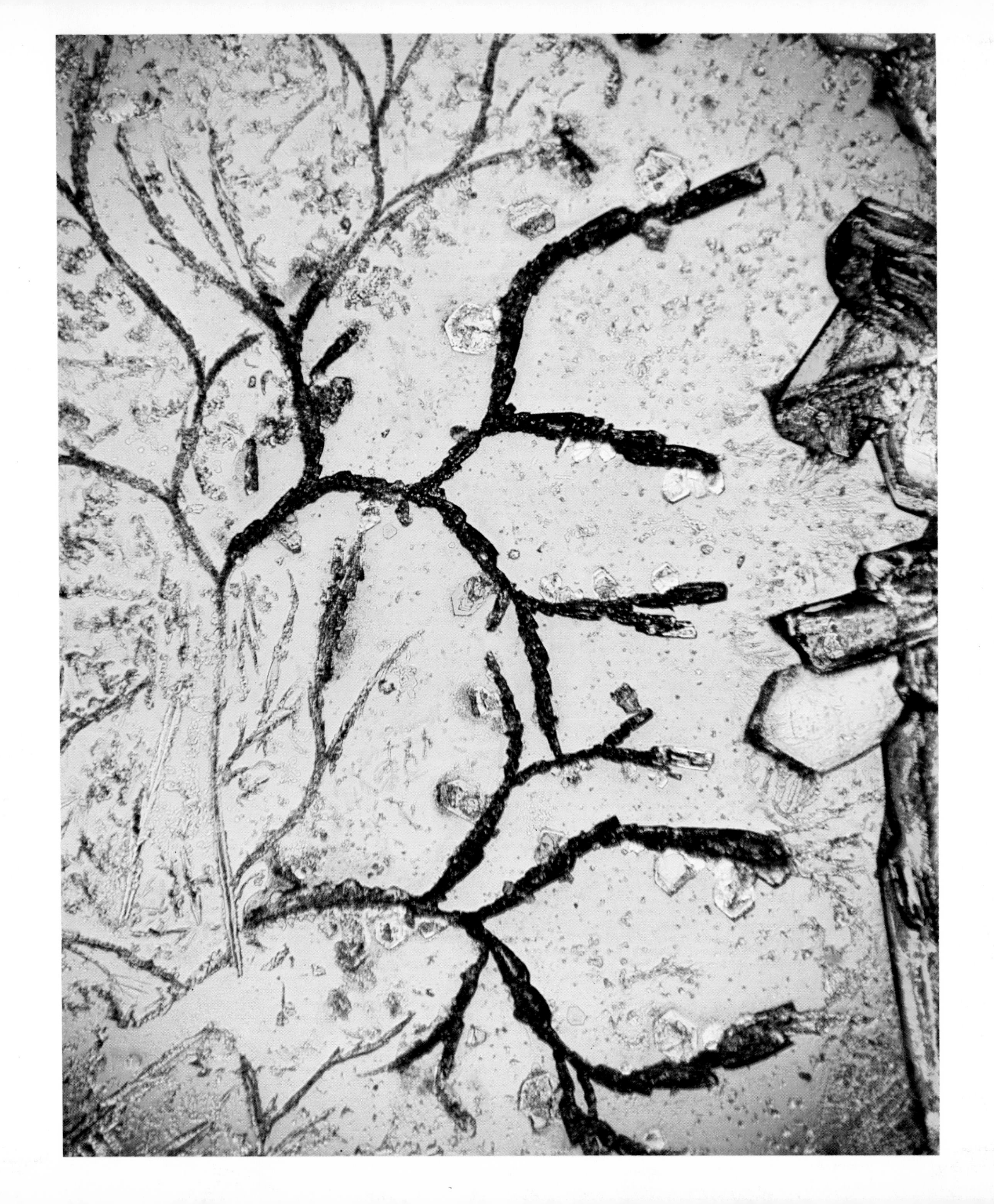

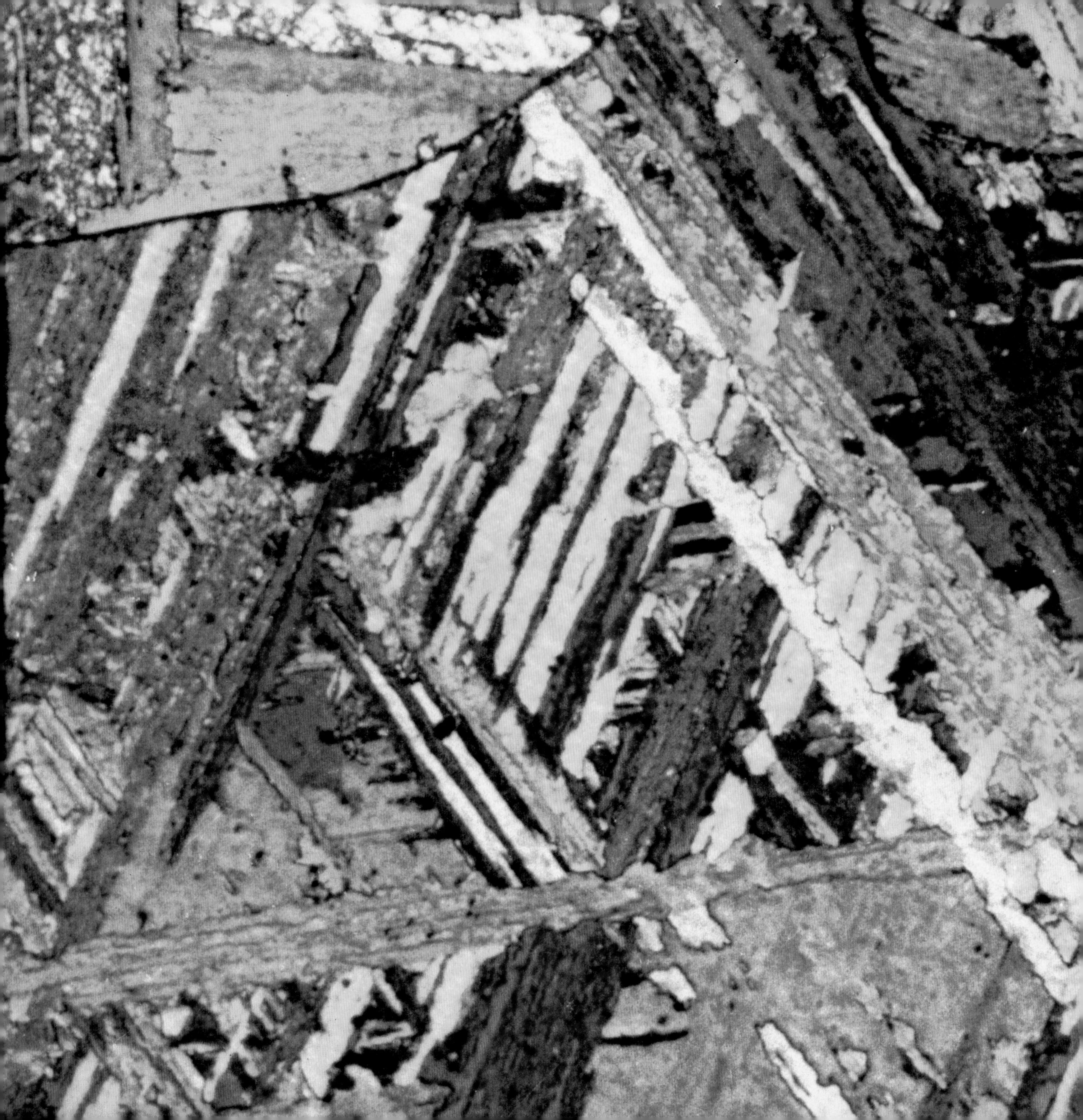

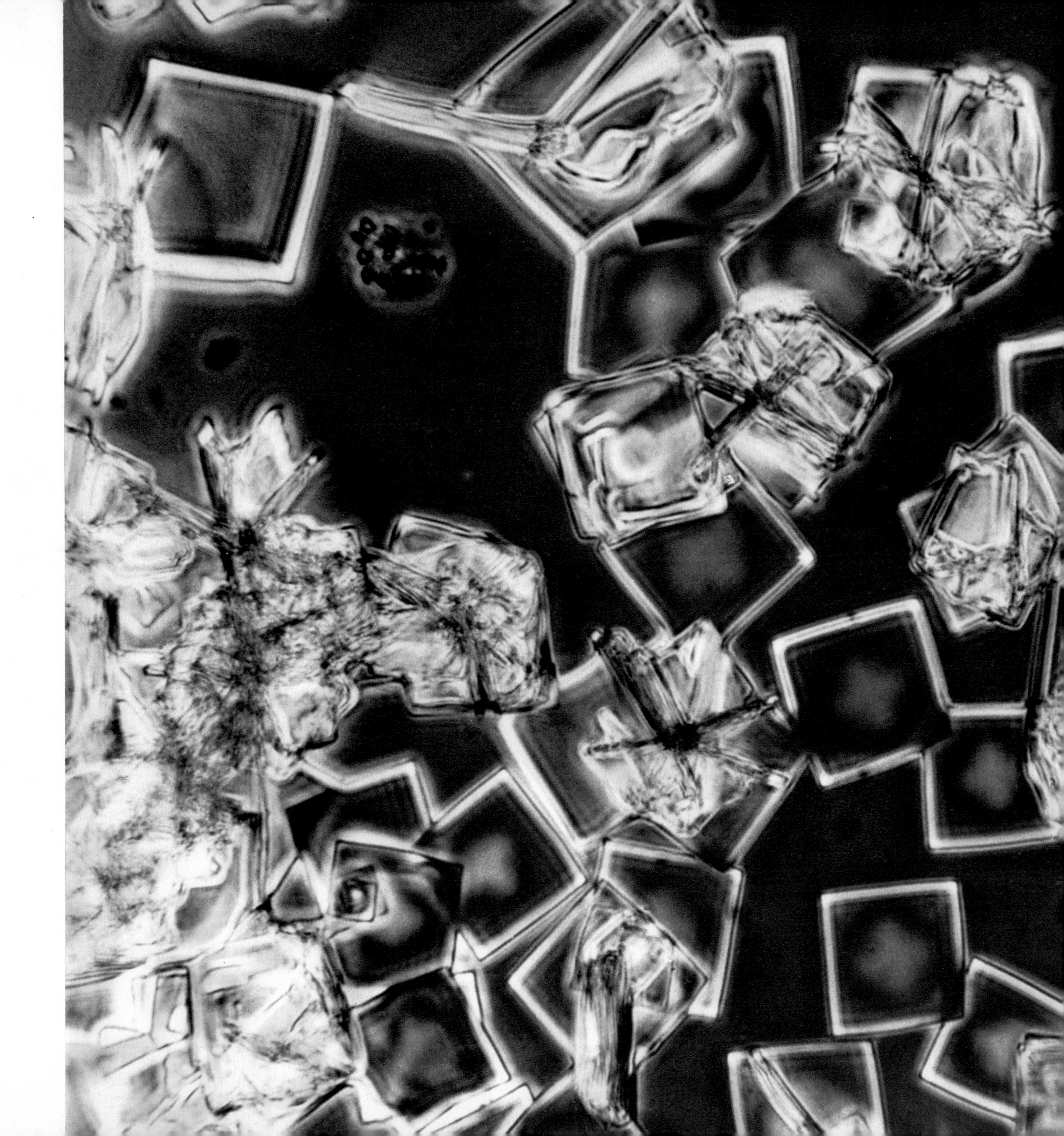

252

253

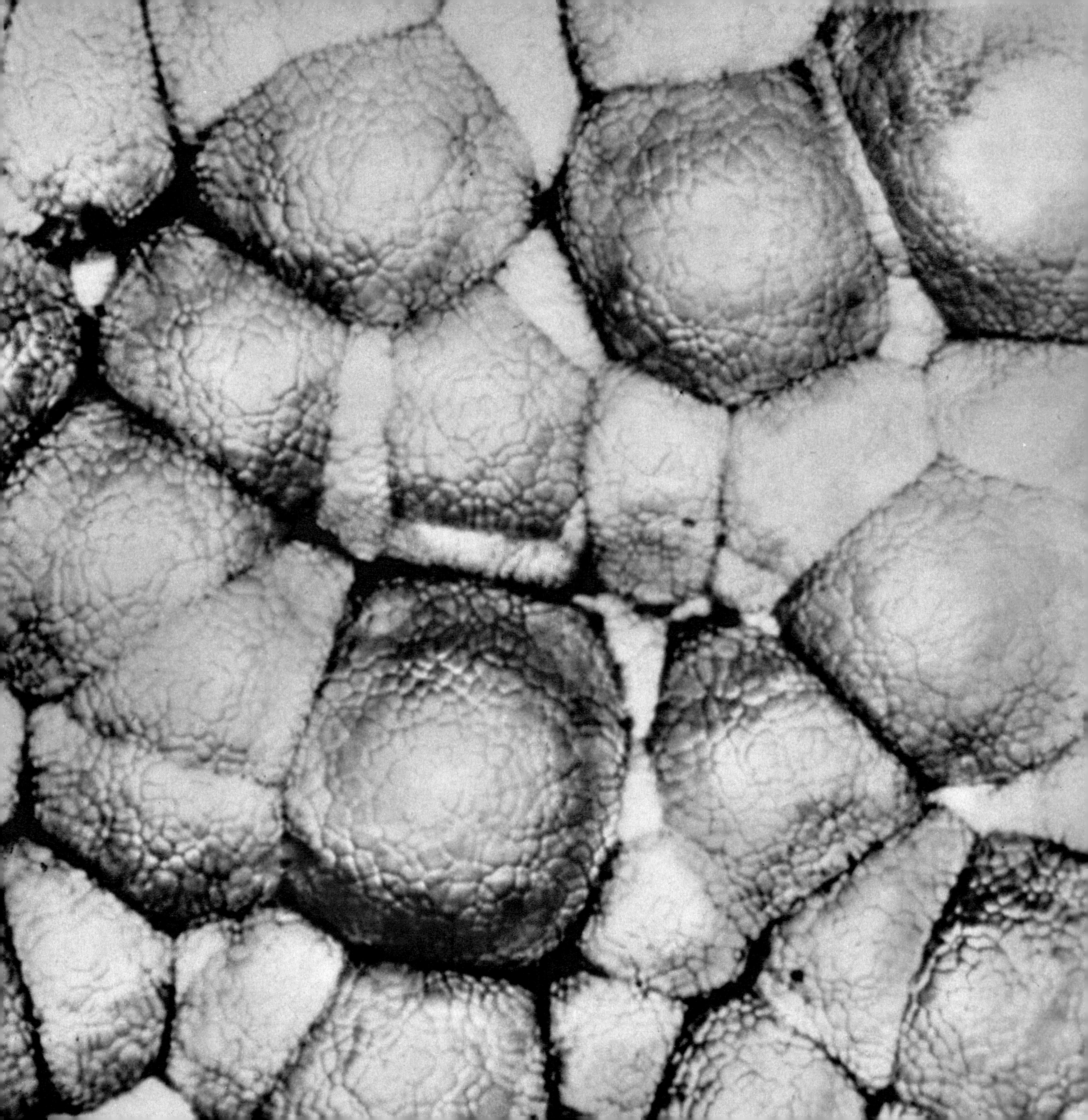

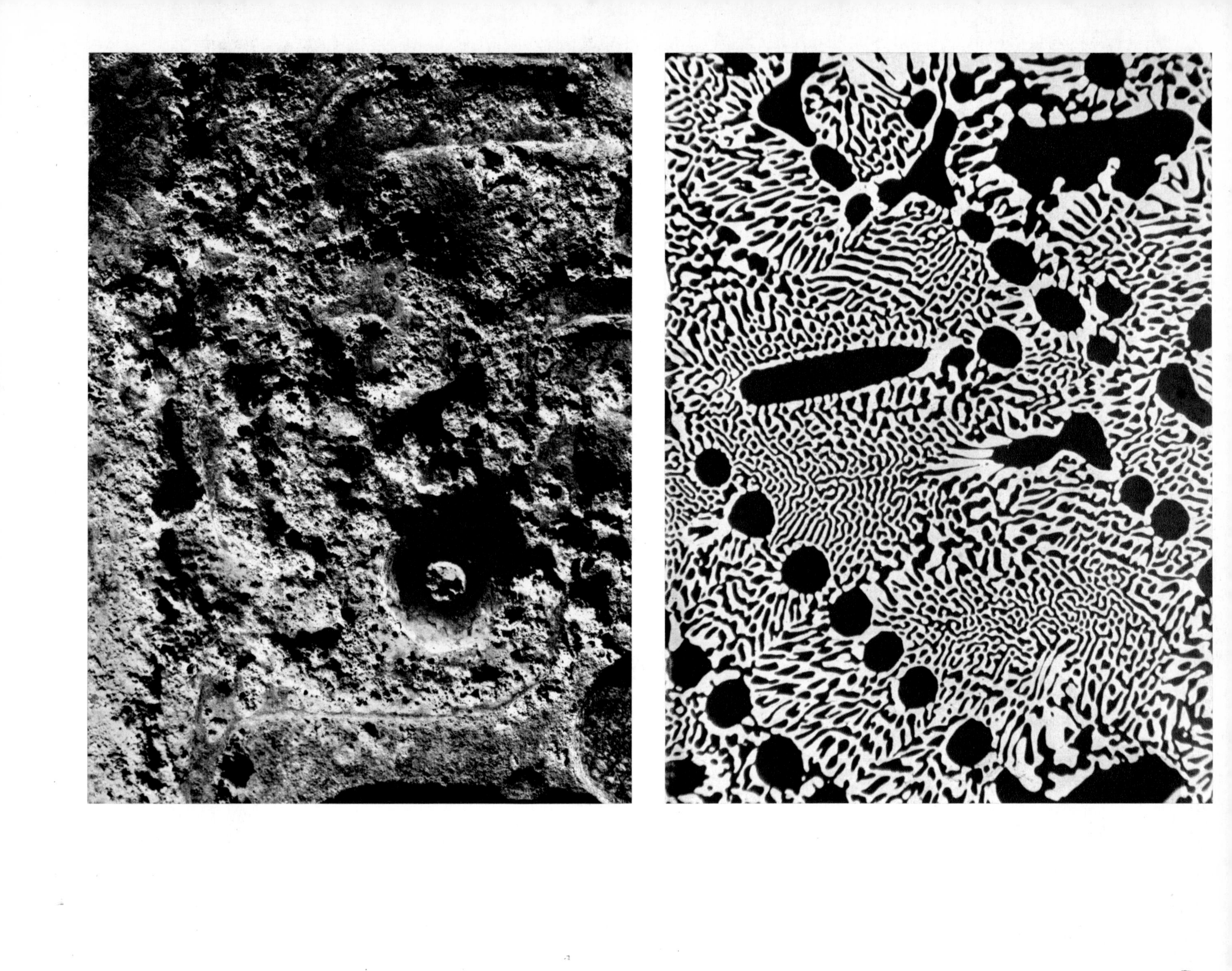

263. Nylon weave
250x

264. Down feather
300x Polarized light

265. Silk threads
375x Polarized light

266. Adherent cellophane adhesive tape
3,000x

267. Cellophane adhesive tape
1,200x

268, 269. Mustard
900x Polarized light

270. Cotton
900x Polarized light

271. Letter *e*
300x

272. Hand cleaner
250x Polarized light

273. Phisohex cleaner
400x Polarized light

274. Lens paper
750x Polarized light, reflected illumination

275. Wool
300x

276, 277. Silk
400x Polarized light

278. Bread (stained)
1,200x Polarized light

279. Borax crystals
700x Polarized light

280. Household dust and debris
300x Polarized light

281. Laundry detergent
375x Polarized light

282. Printing on 1¢ postage stamp
1,000x Dark field, reflected illumination

283. Metal staples
180x Reflected polarized light

284. Toothpaste
160x Polarized light

285, 286. Sugar crystal
1,900x Polarized light

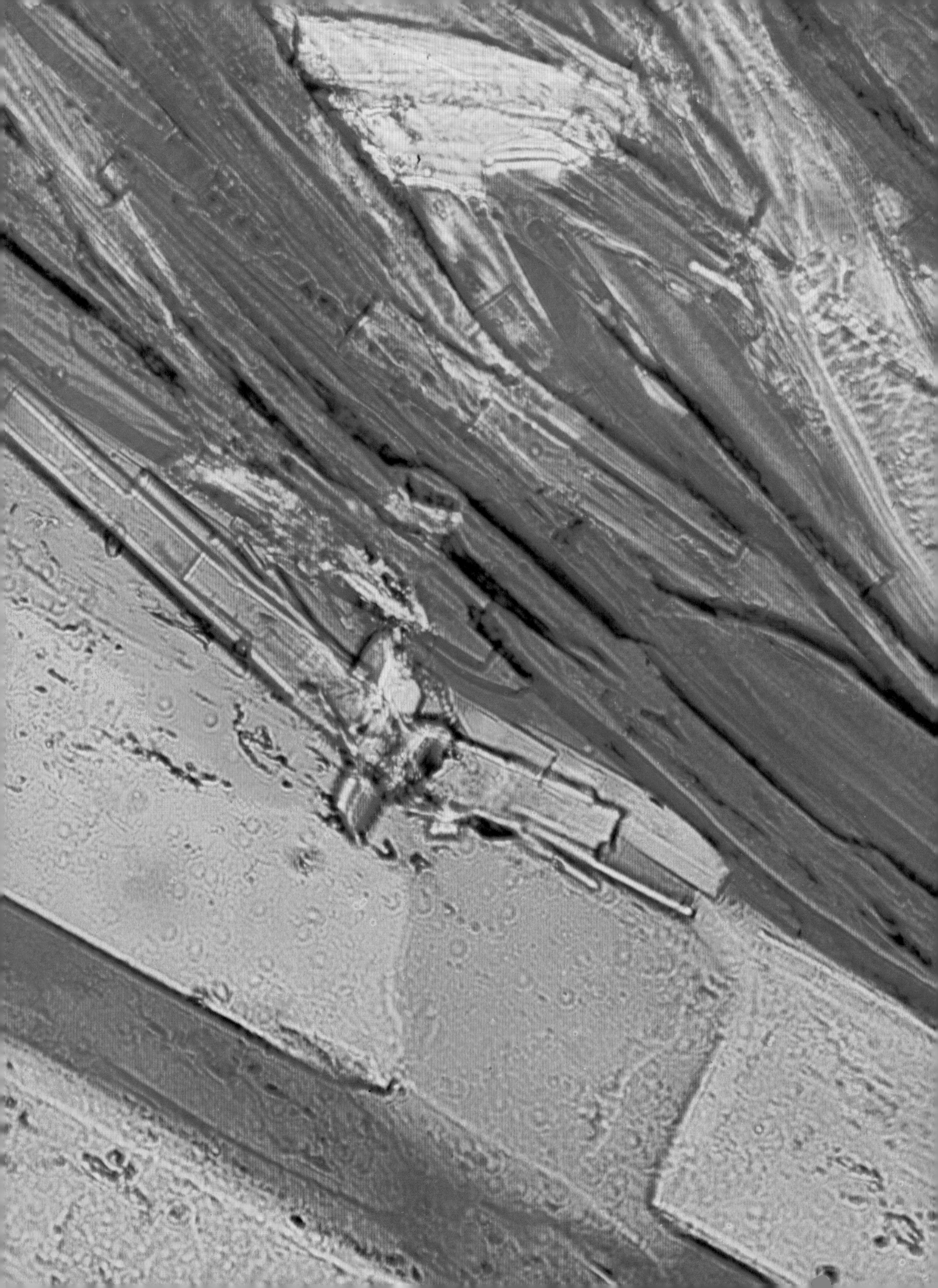

FAMILIAR OBJECTS

Keep fold-out (at left) open for captions to the following section

Obviously, this final section represents a nonscientific category, cutting across many of the classifications illustrated in the previous sections. These photomicrographs are assembled purely for the enjoyment of the reader who is curious to know what sorts of microscopic forms reside in his own household—particularly in the everyday objects that surround him. The panorama of microscopic patterns, shadings, and tints hidden all around us may strike the viewer as a poignant demonstration of esthetic universality.

There is no dearth of subjects. A wood chip from broken furniture, the frayed edge of a mop, the hair of pets, pillow stuffings, leftovers from the salad bowl, even household dust—all take on a mantle of incredible beauty under the microscope.

A bit of fabric from clothing, draperies, rugs, or linens—when slightly teased, positioned on a slide, carefully saturated with a fixative, and topped with a cover glass—will often be transformed into inspired luminous interlacings. The weavings of homespun, rep, and twill have unique motifs. Animal fibers (alpaca, mohair, cashmere), vegetable fibers (jute, hemp, sisal, ramie), mineral fibers (asbestos), glass fibers (fiberglass), metal fibers (Lurex), synthetic fibers (rayon, nylon)—all display distinctive structures. For example, raw silk has a continuous double thread with adherent clumps of fiber scattered throughout; treated silk is a filament with no details; cotton fiber is flat and twisted; kapok is seen as a thin-walled cylinder with air bubbles.

Different kinds of paper viewed under the microscope yield fascinating images; for instance, papers derived from wood pulp reveal pits and spirals in areas where living cells once existed. Printed and engraved items often resemble majestic abstract works of art. A tiny area on the tip of someone's nose in a color photograph turns into a stunning arrangement of distinctive dots of primary color—which easily could be taken for modern painting. The medicine cabinet is a cornucopia of potential specimens, which—with minimal preparation for the microscope—explode into stunning arrangements of crystal faces and cleavages.

With a bit of ingenuity the amateur, if he is fortunate enough to have a microscope, will find a lifelong resource of specimens in the art gallery of his own home.

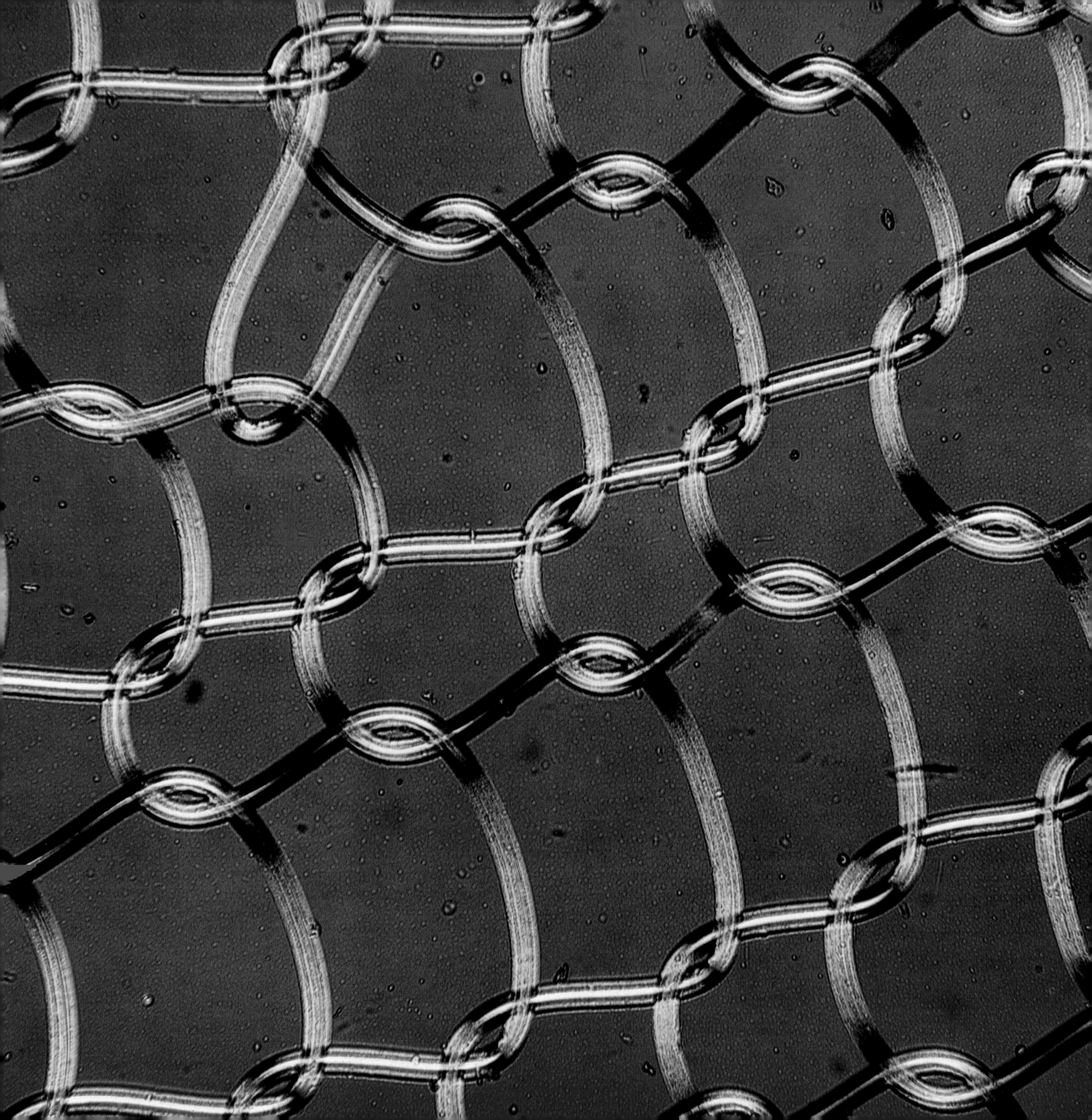

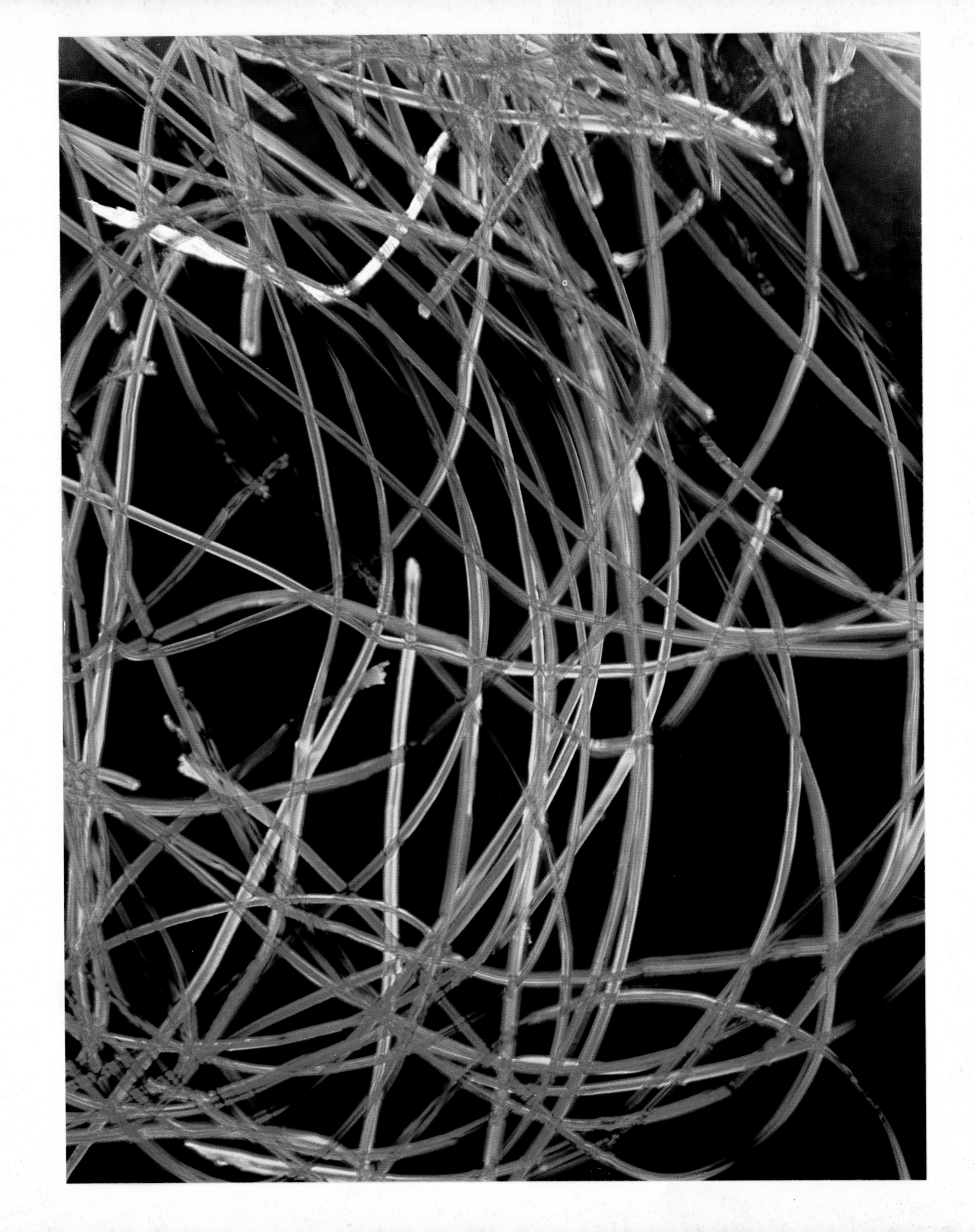

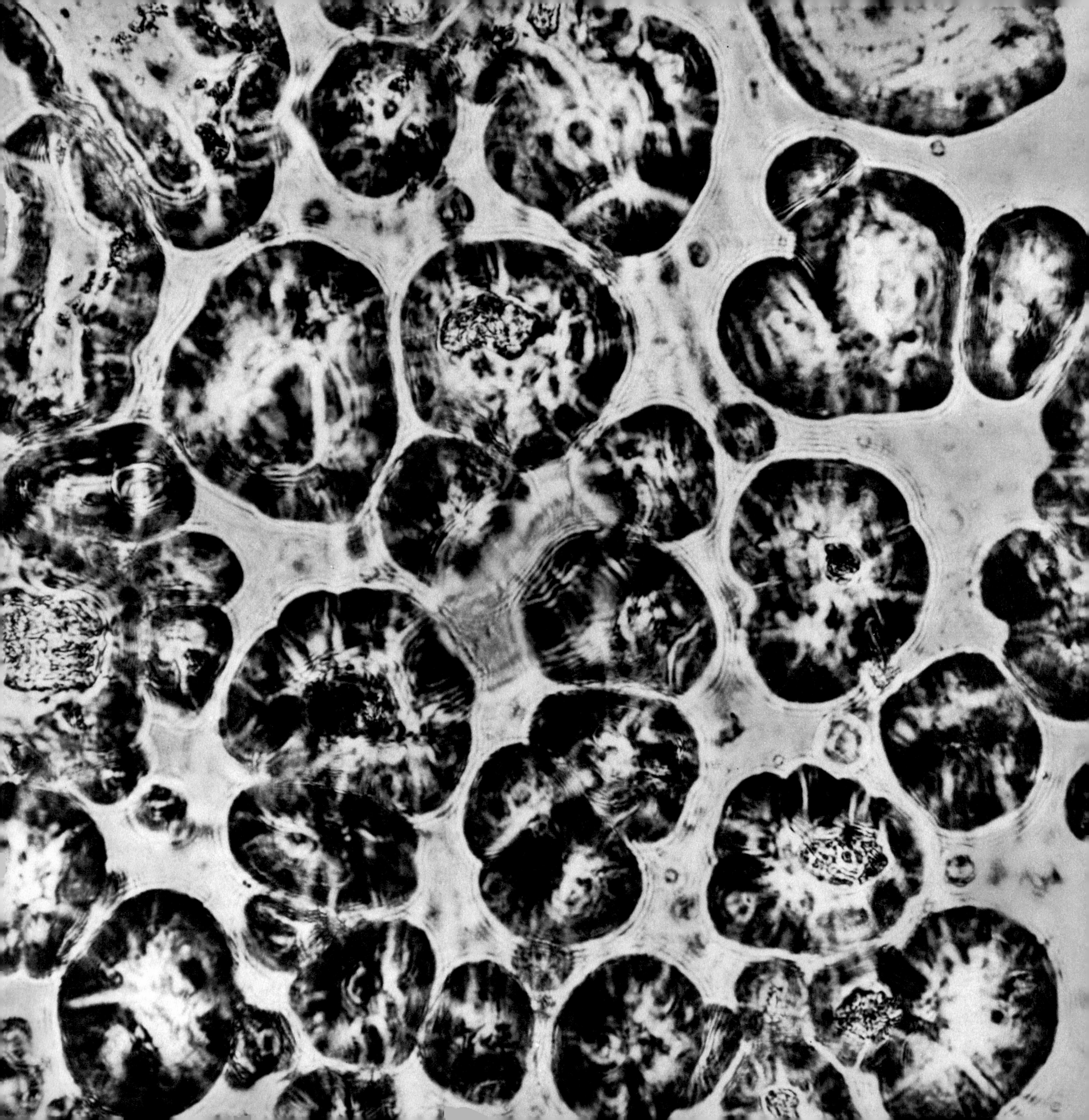

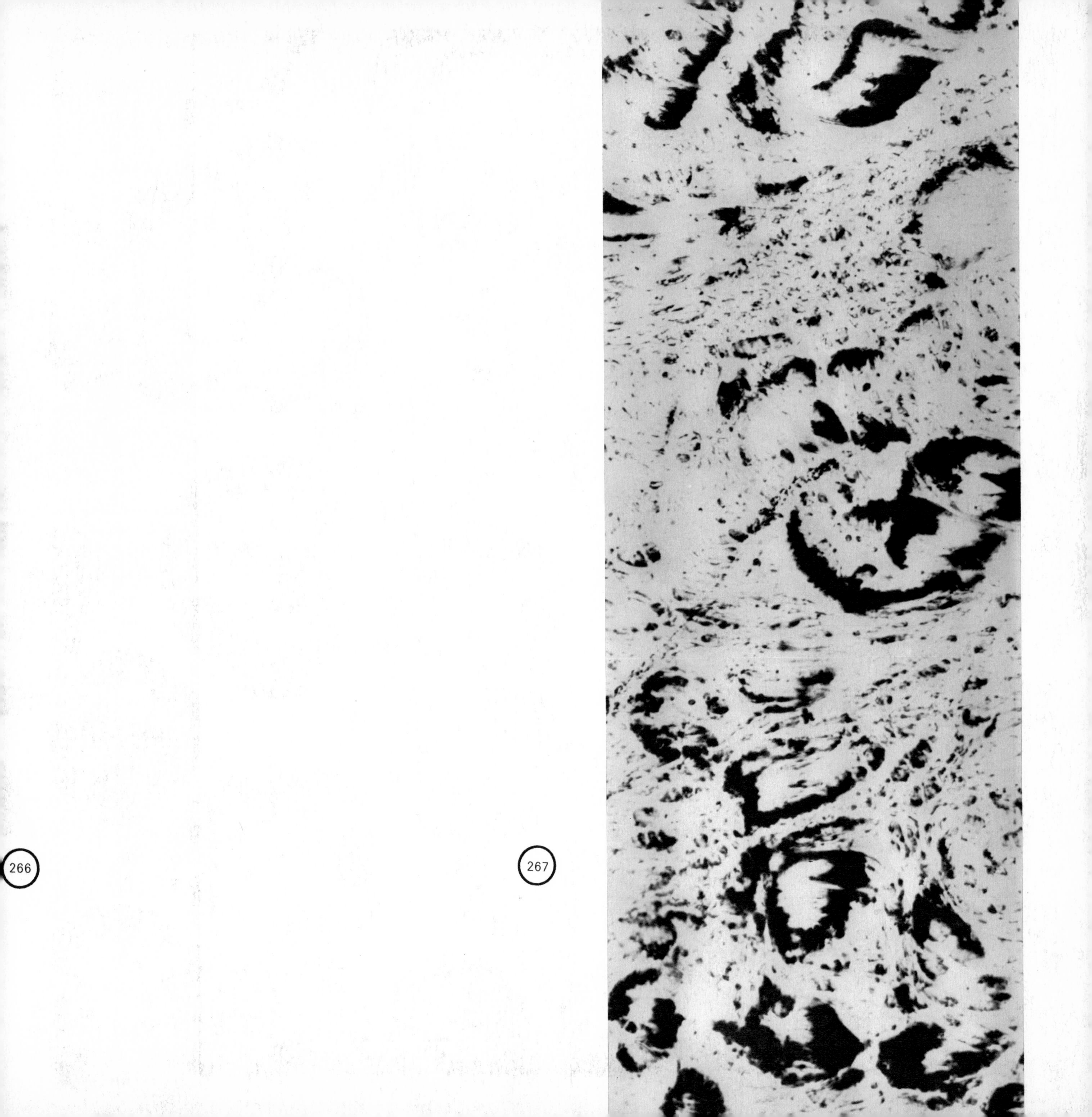

266

267

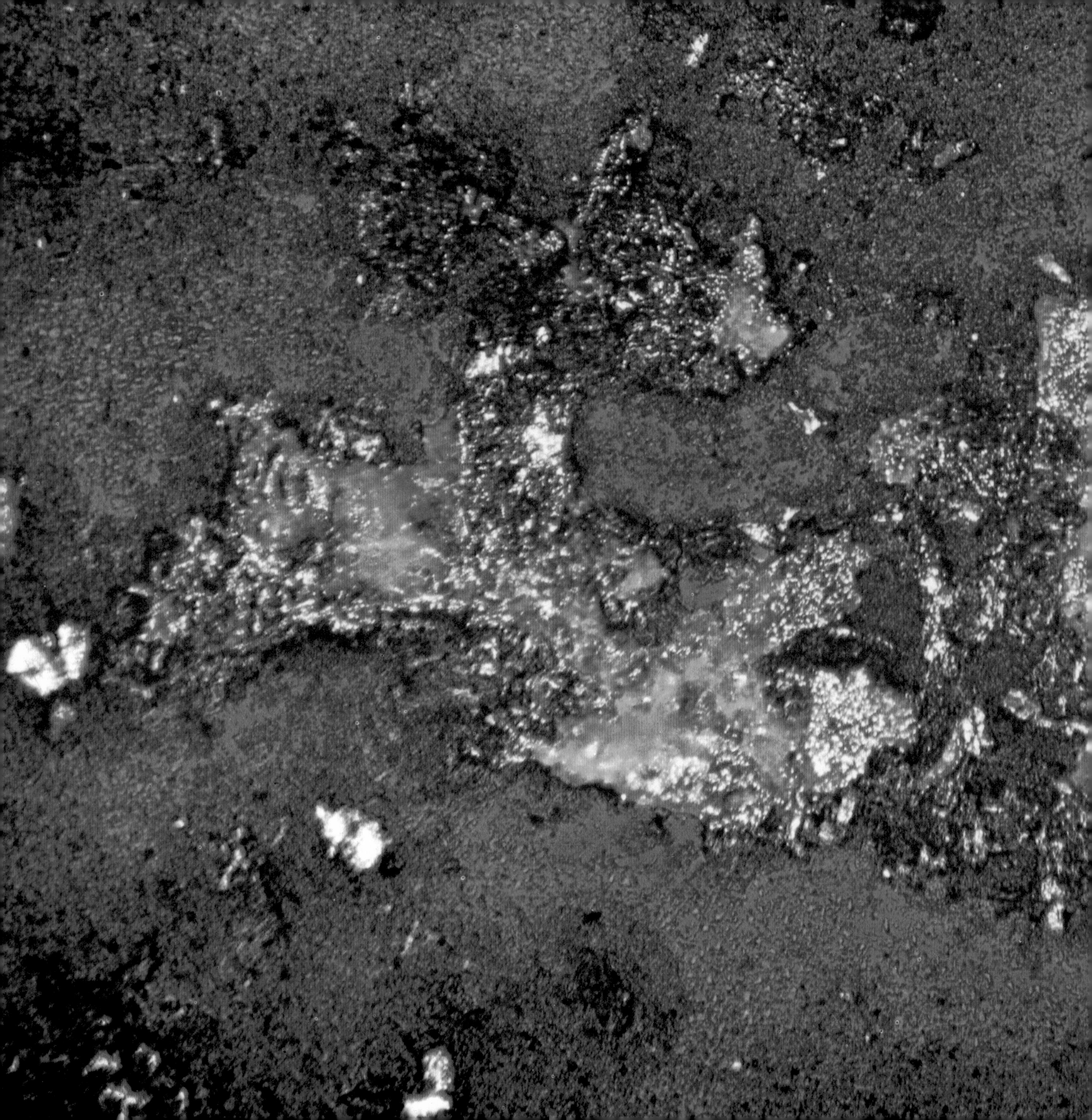

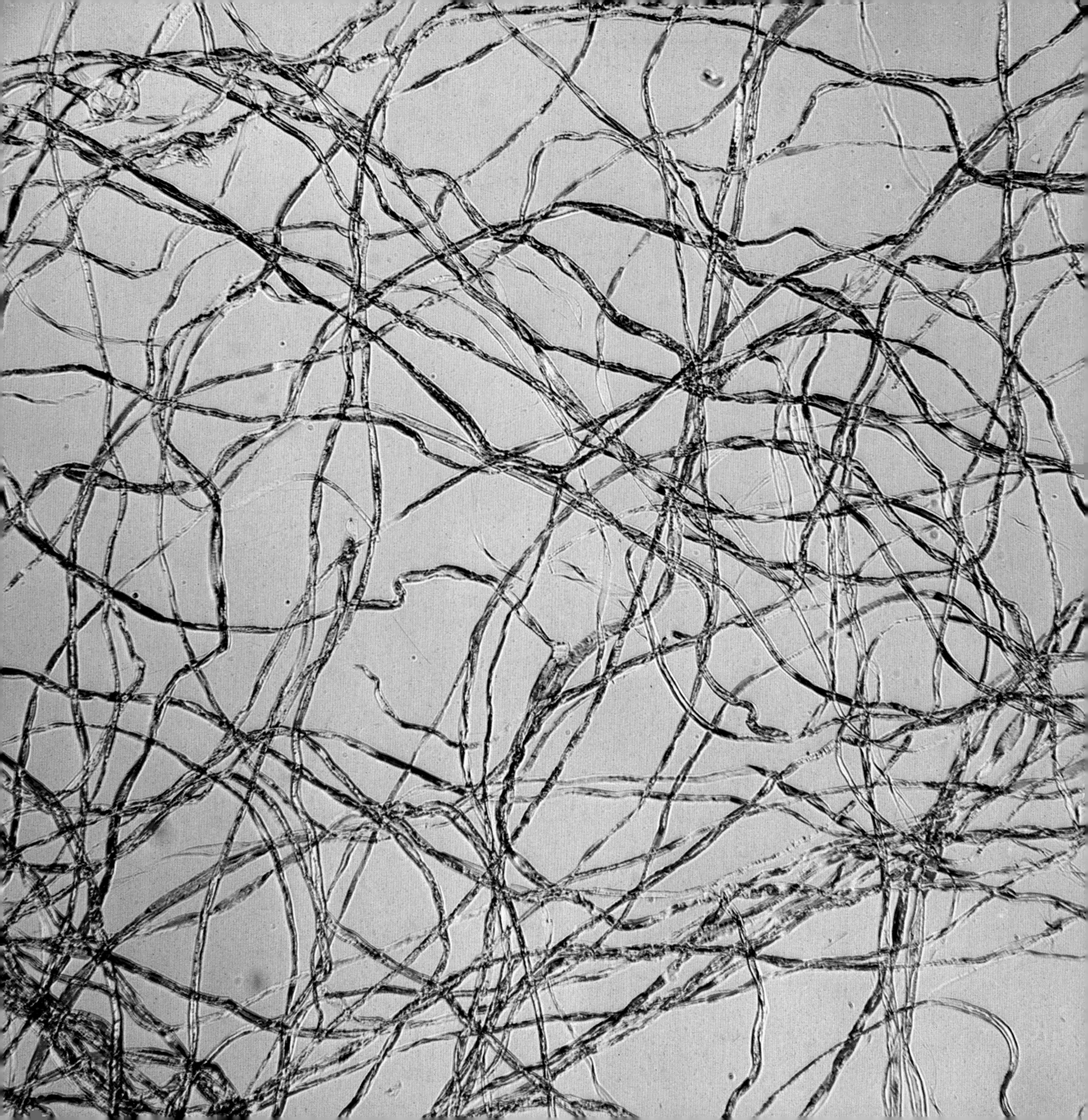

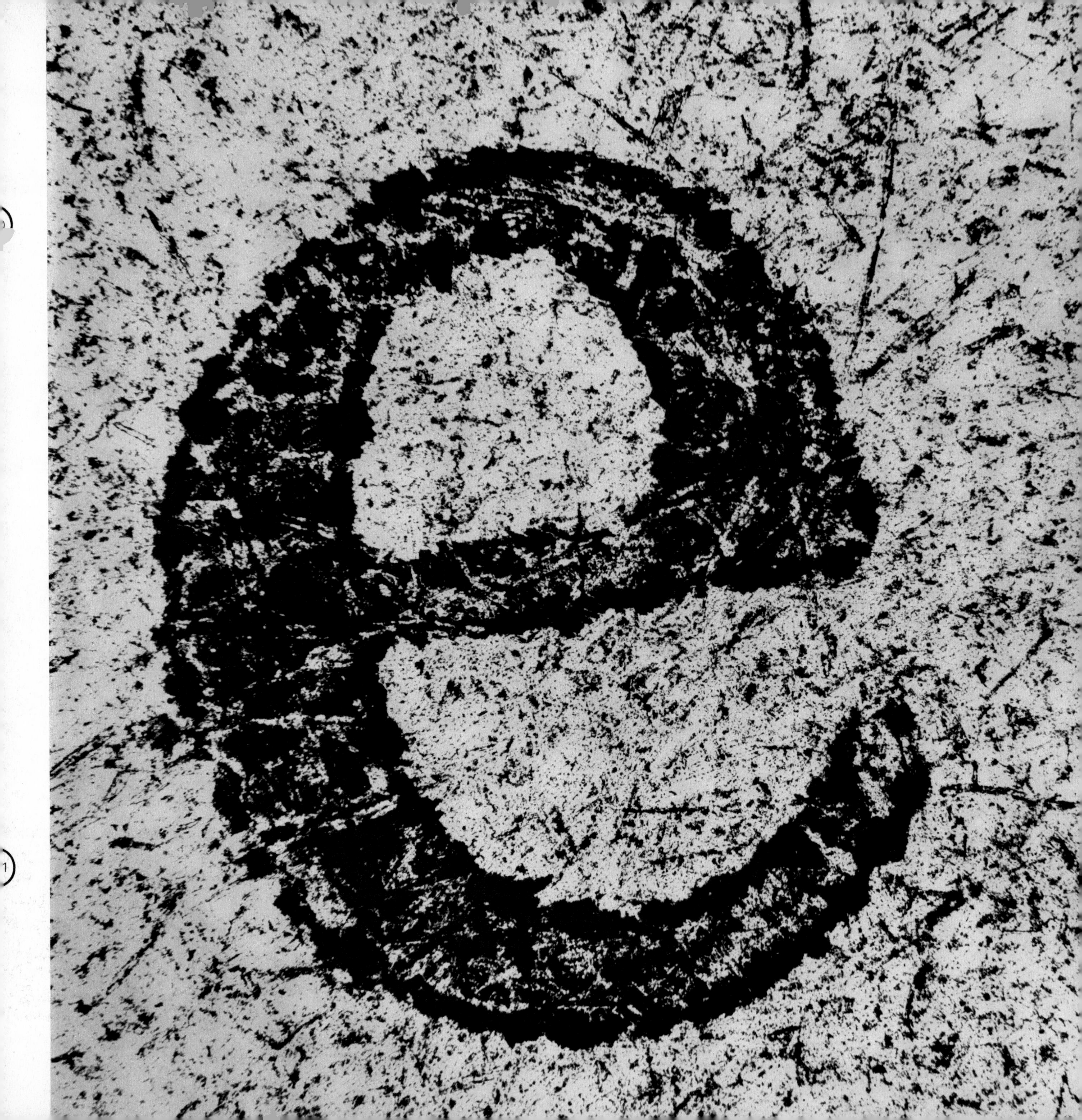

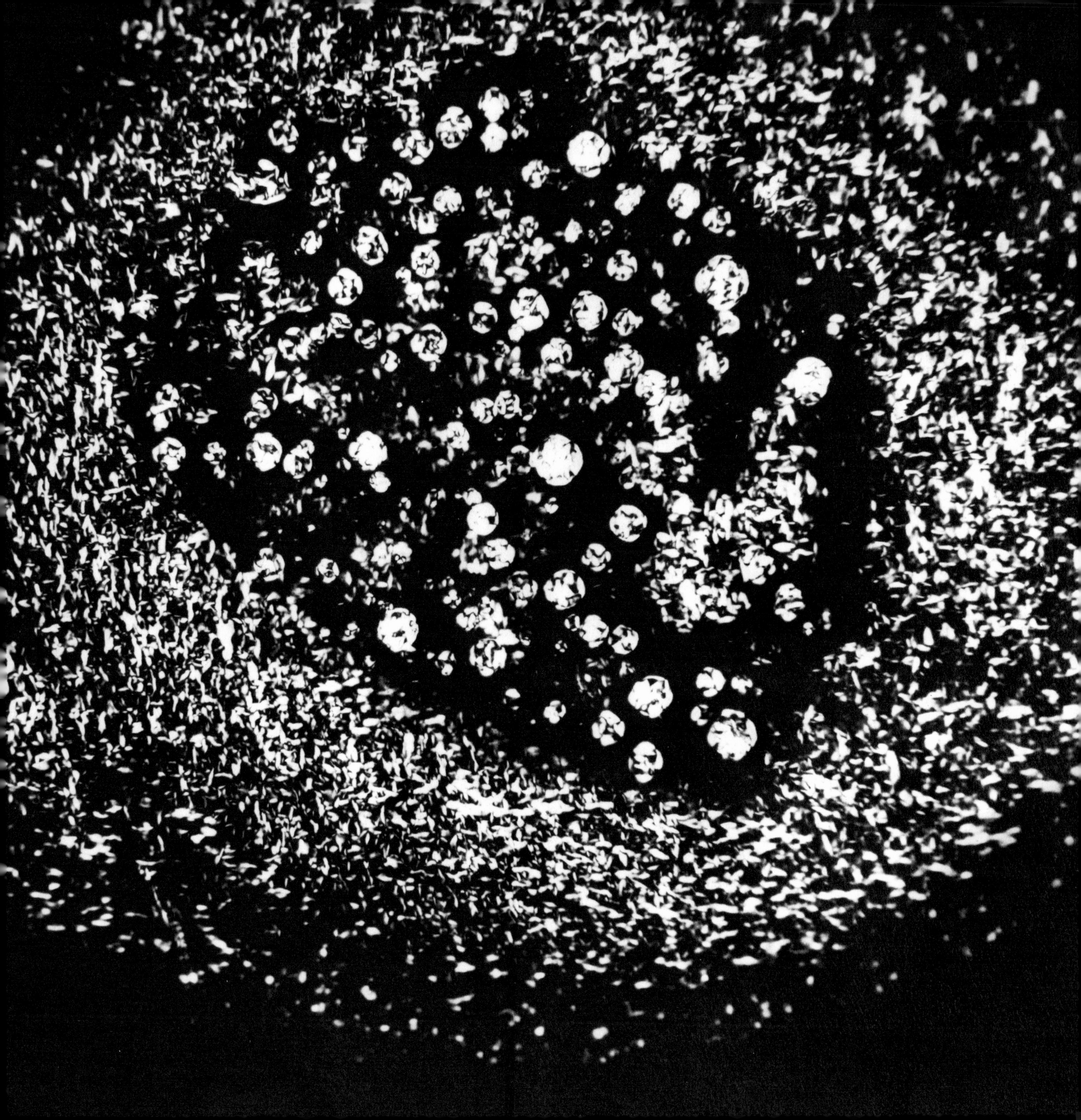

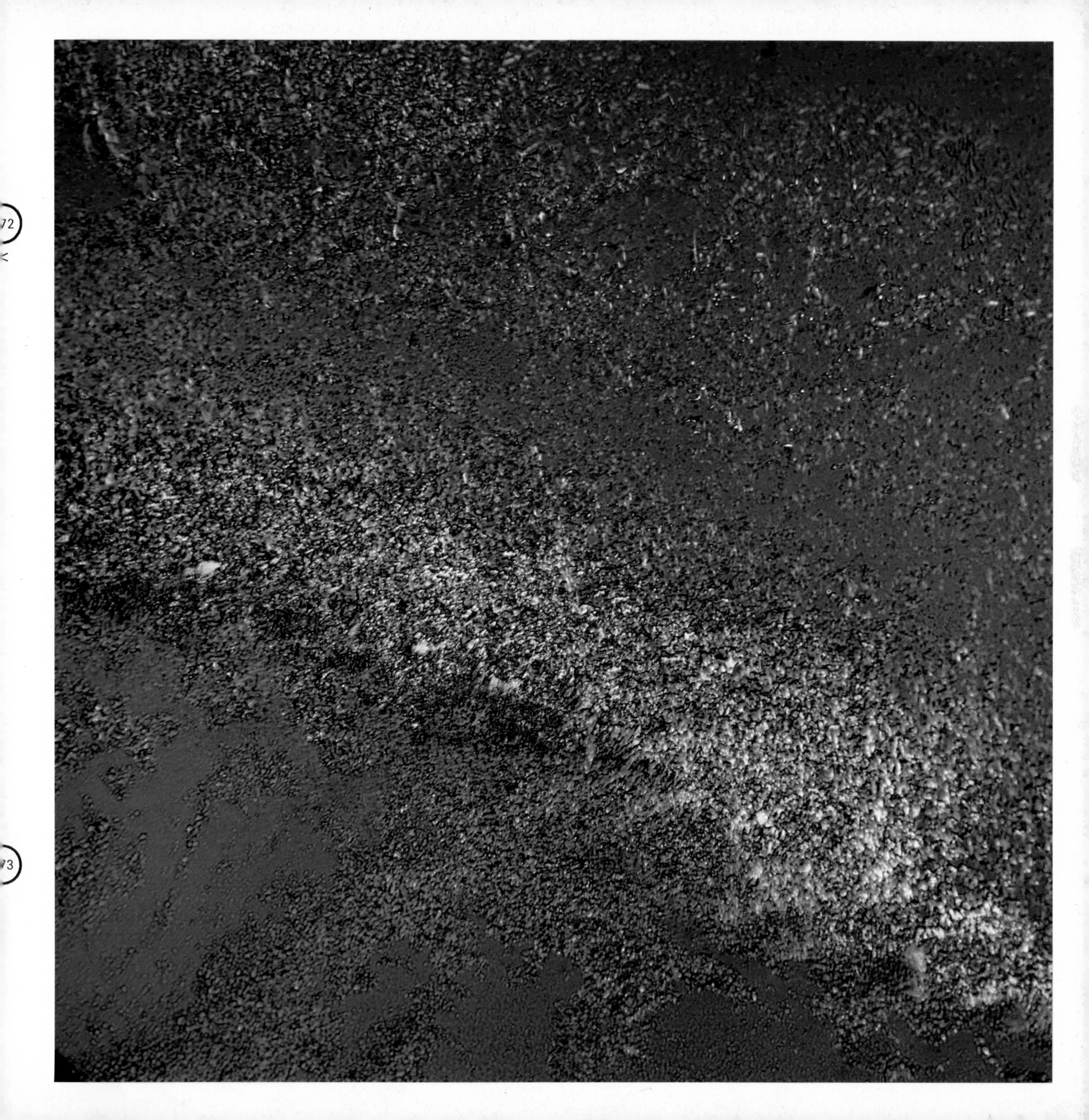

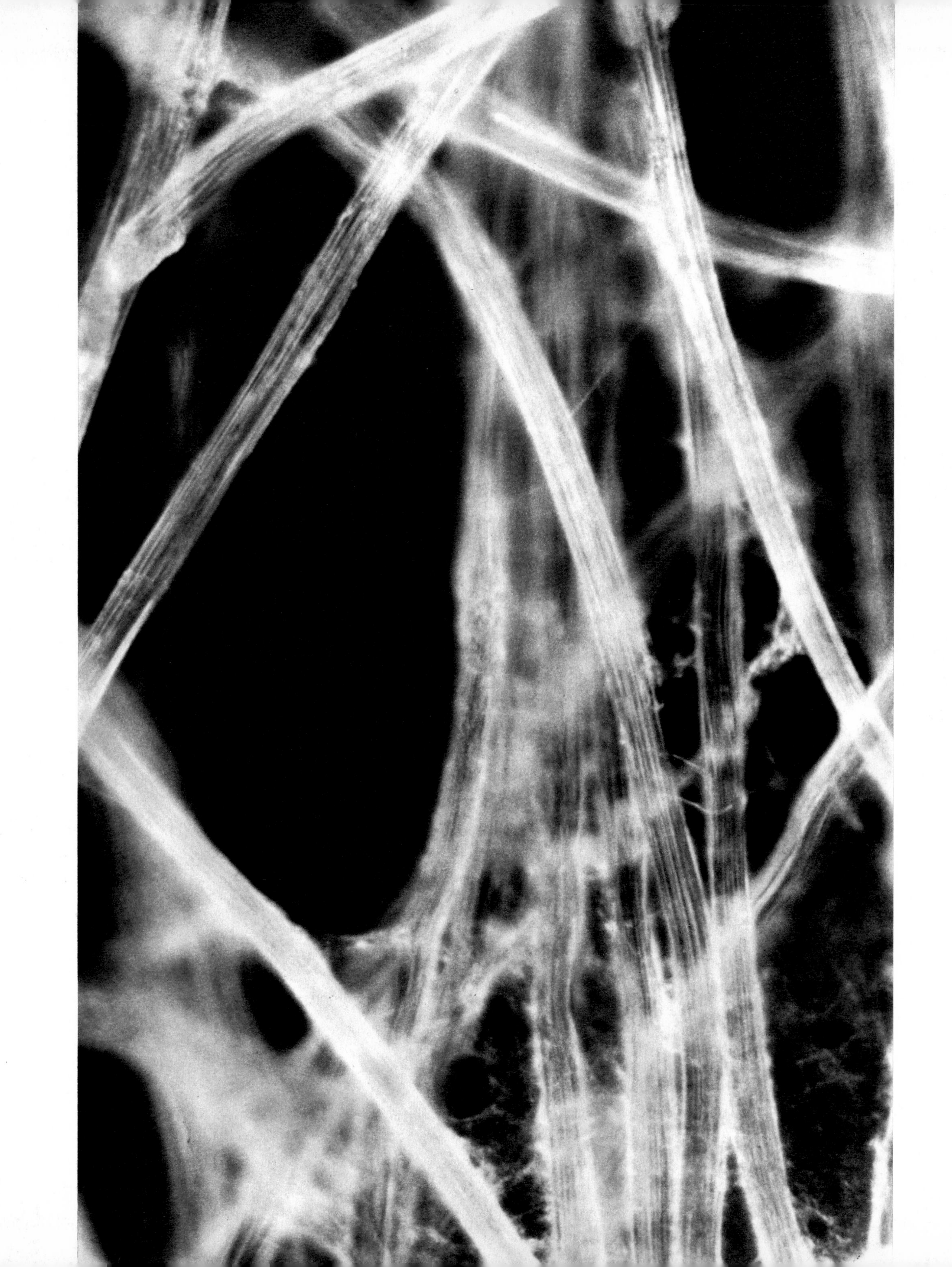

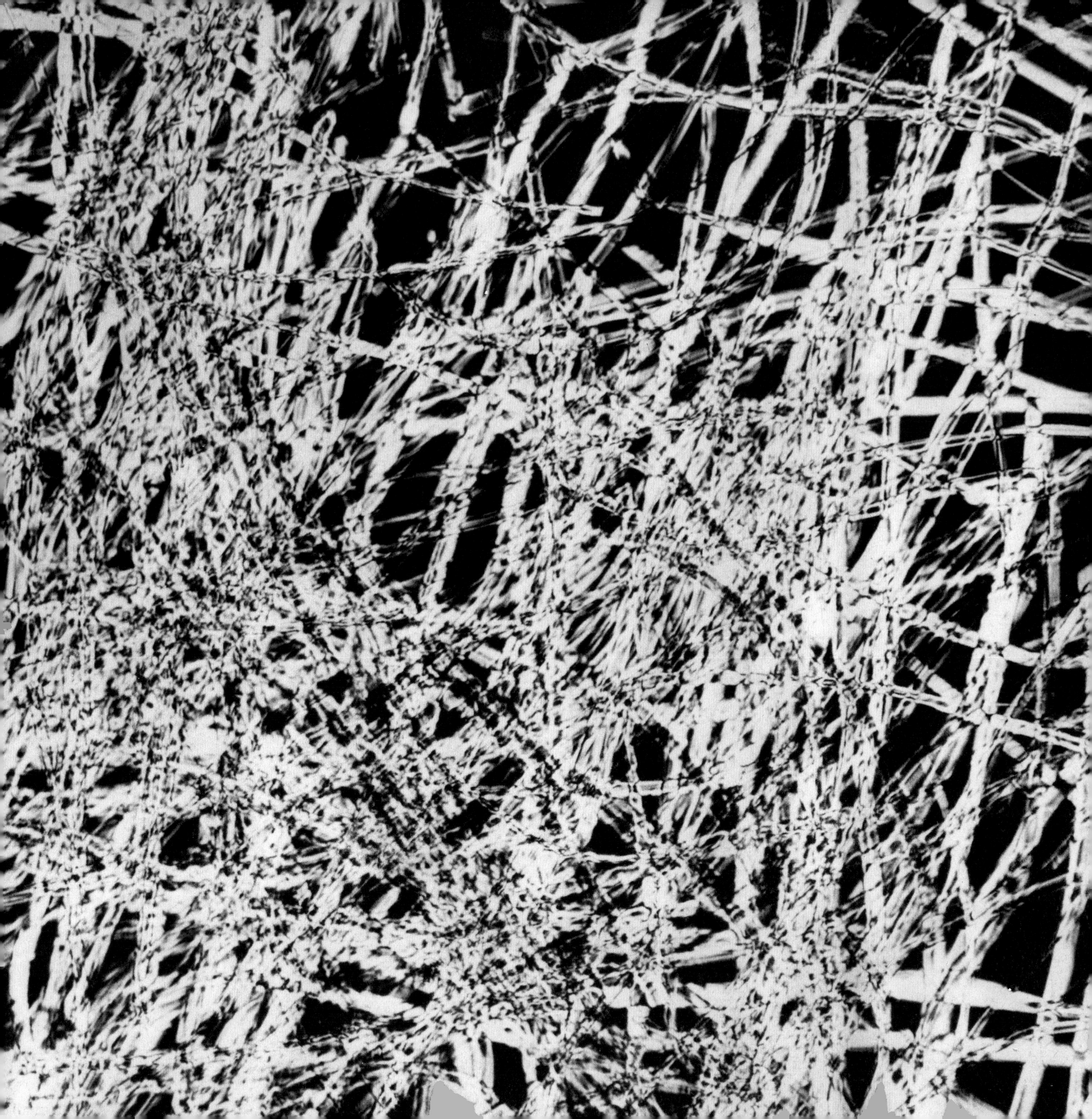

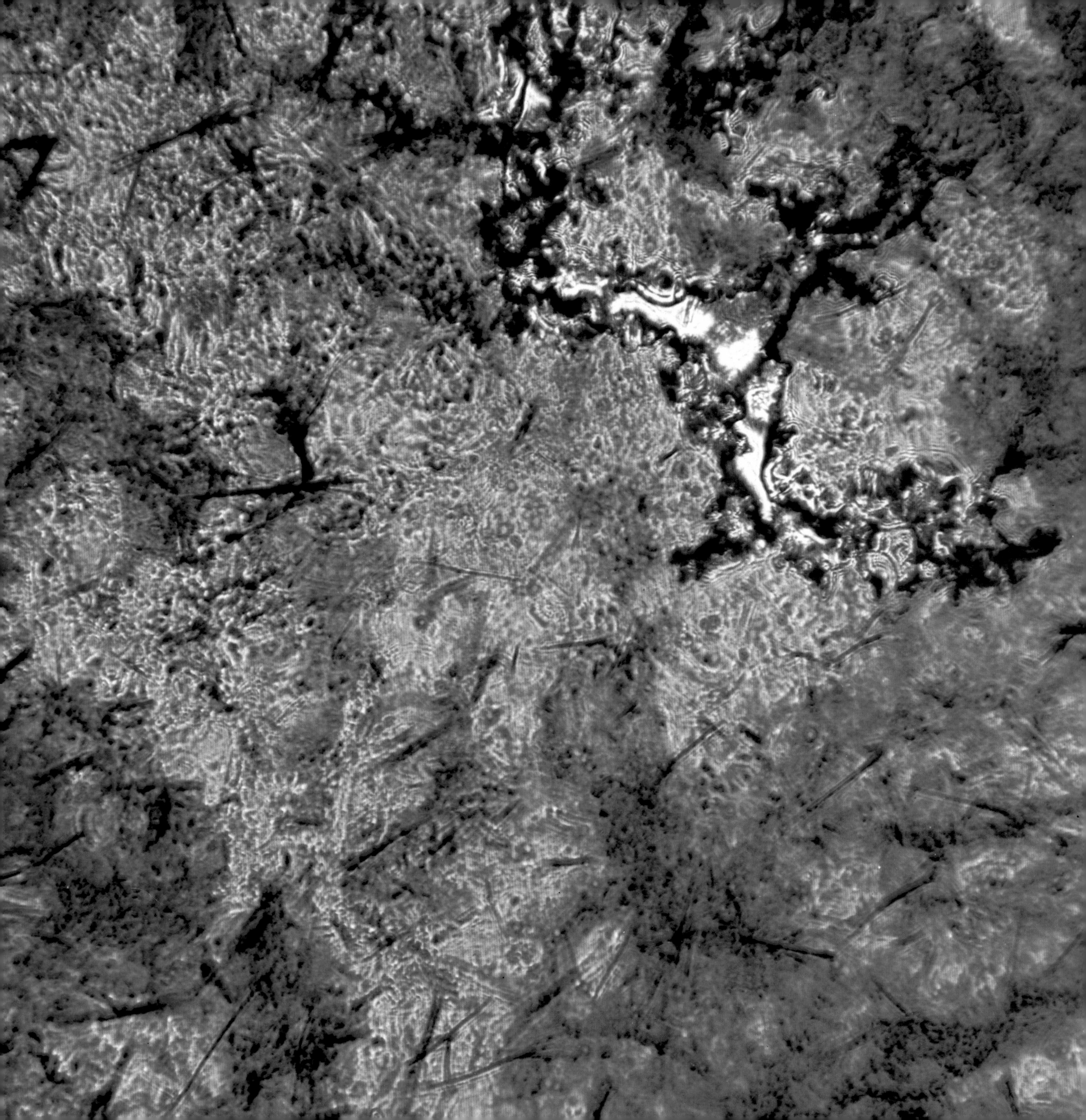

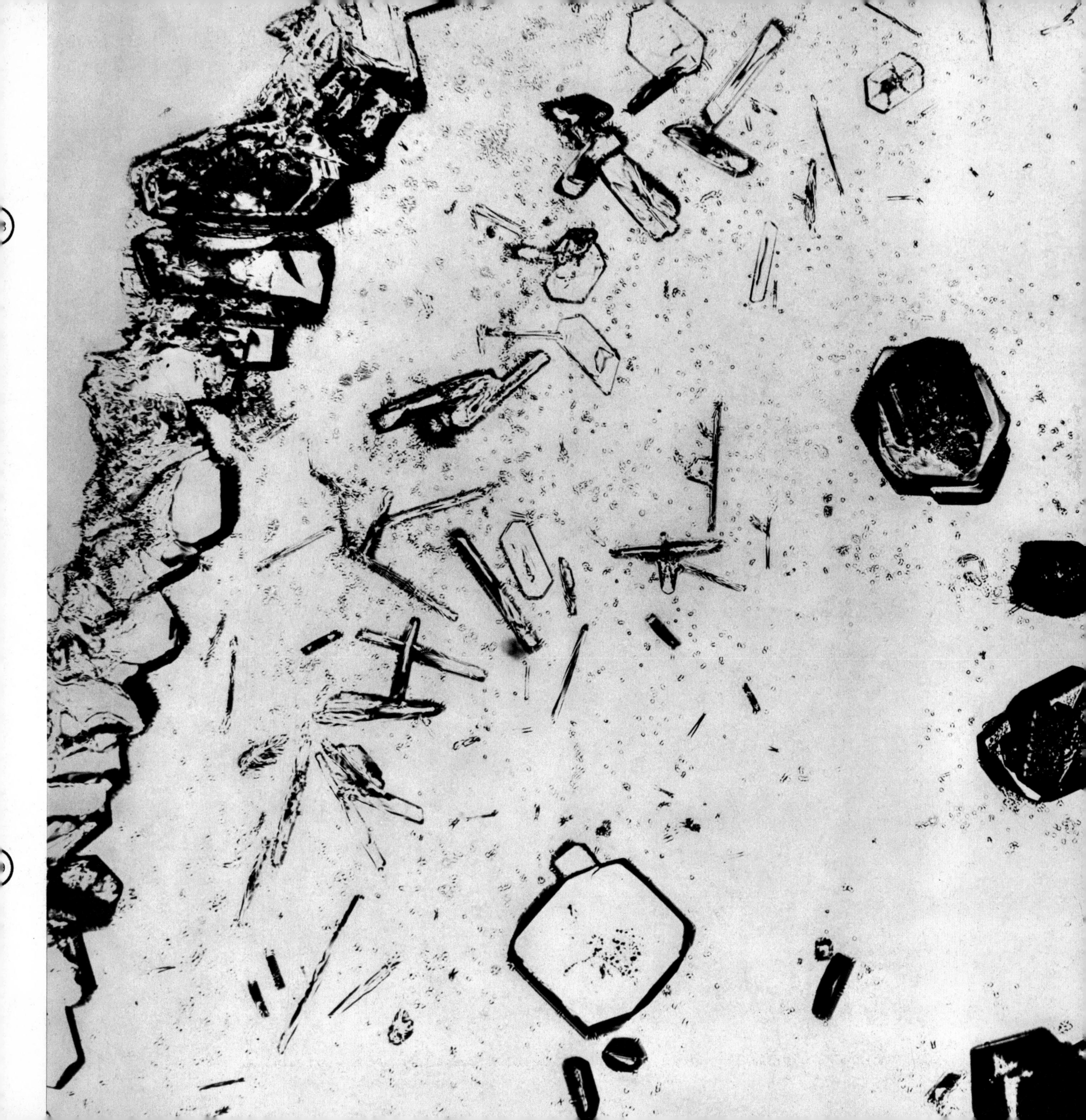

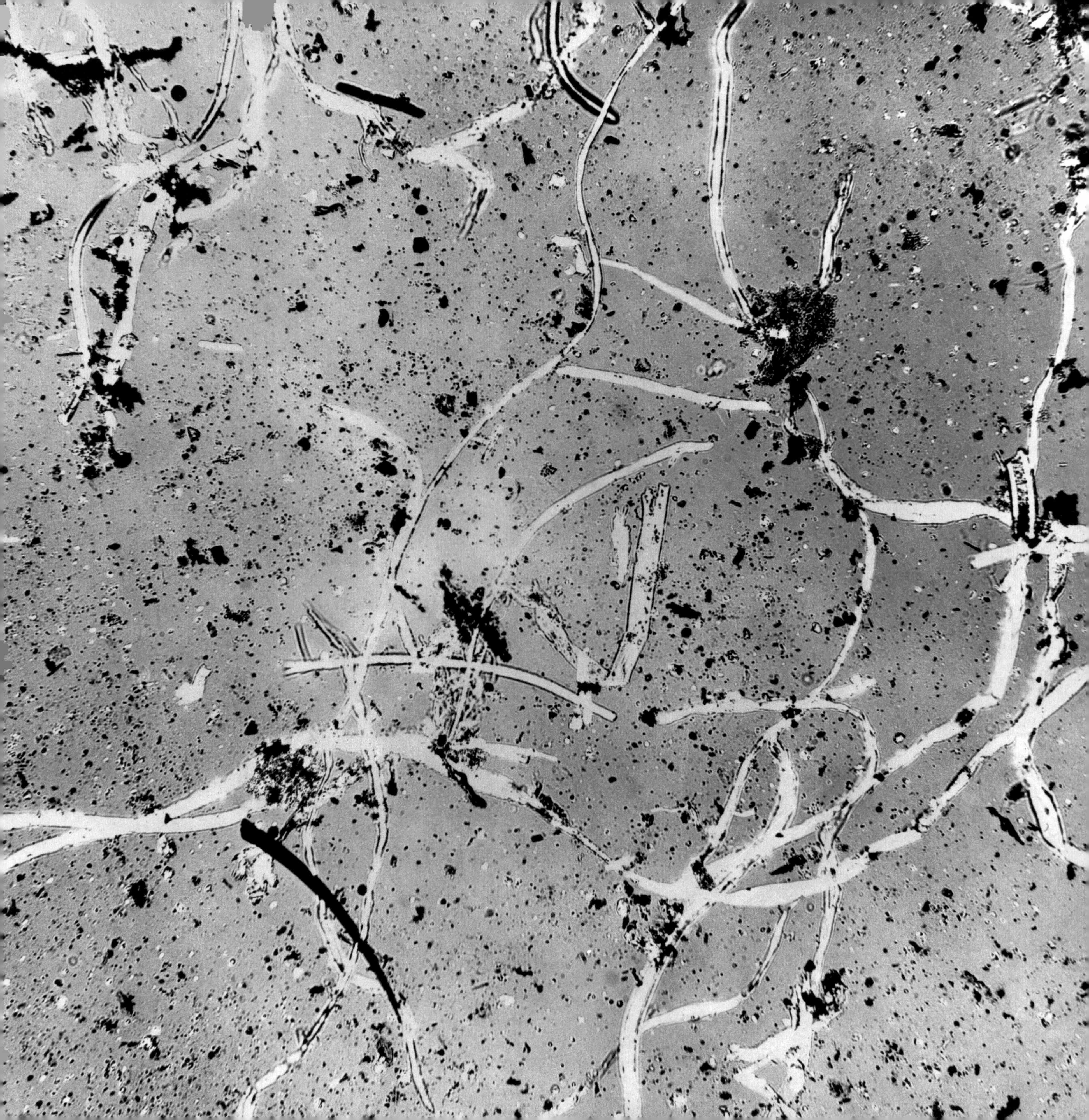

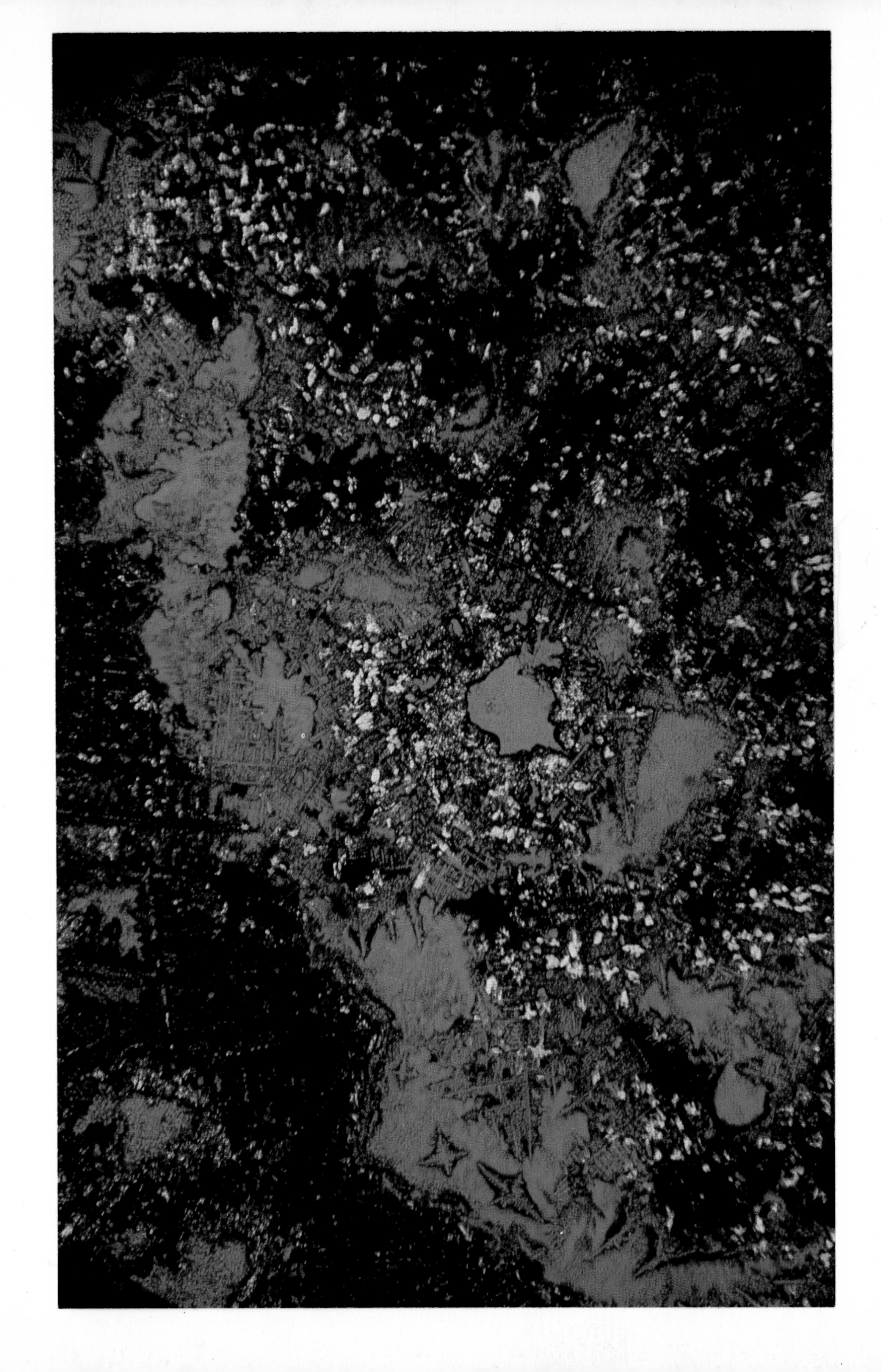

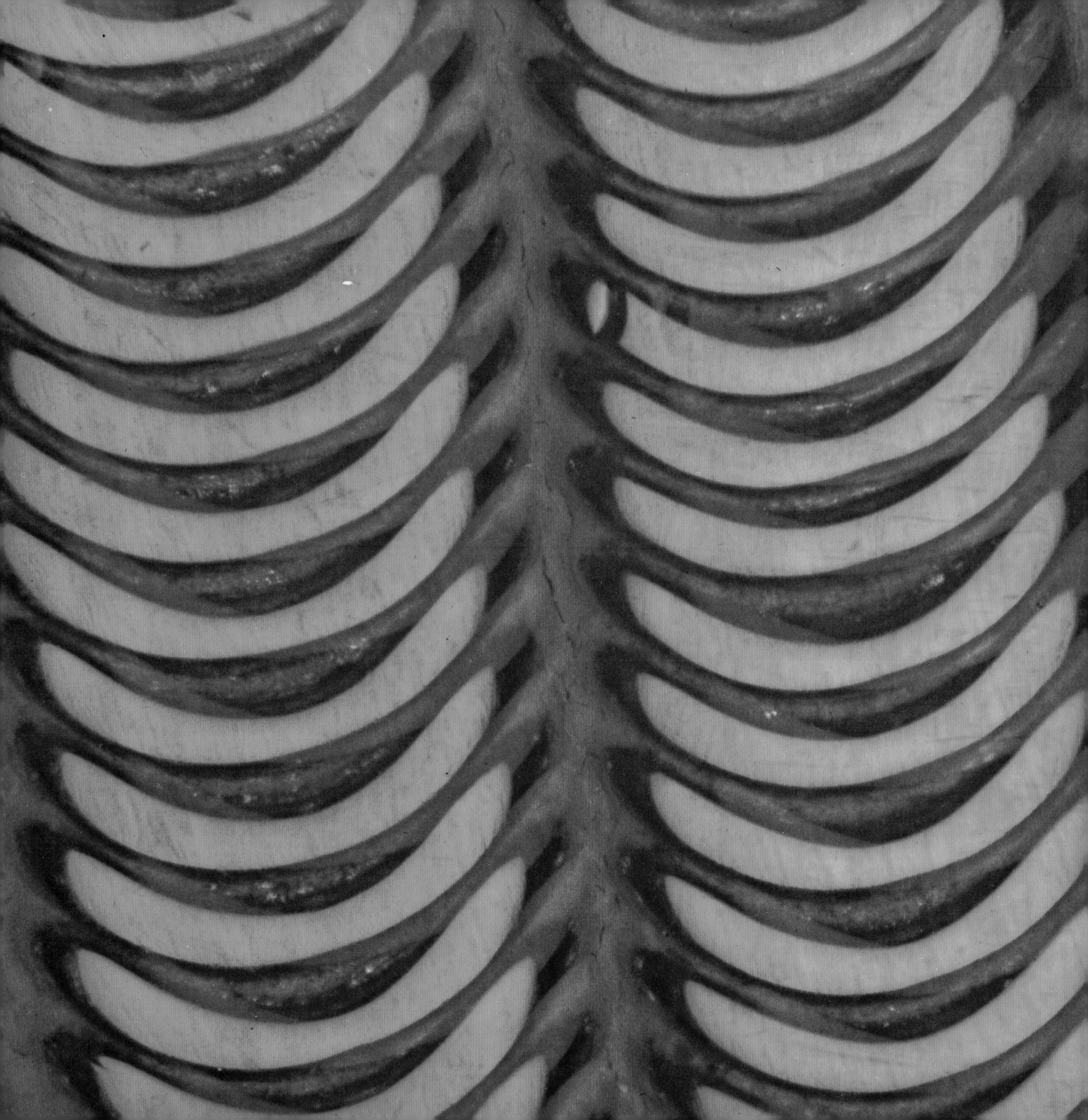

APPENDIX

Photographic equipment and film

The camera most often used for the illustrations in this book was a Rolleiflex SL 66, a single-lens reflex camera, equipped with a prism-focusing device attached to the ground glass. This camera accommodates 120-type roll film and produces a picture size of $2\frac{1}{4} \times 2\frac{1}{4}''$. No accessory shutter attachment was necessary because the camera has a focal plane shutter. The second camera was a Linhof $4 \times 5''$, which uses cut film. A special reflex attachment permitted focusing onto the ground glass. A leaf-type shutter with a microscope adapter replaced the front lens.

All the microscopic equipment was of Zeiss manufacture and included a tri-ocular eyepiece arrangement to facilitate focusing. Most of the photographs were taken through plano-apochromatic objective lenses; the phase-contrast work was done with fluorite lenses; and the vertical illumination and interference microscopy required special lenses known as epiplans. A combination aplanatic/achromatic condenser permitted variable bright-field, dark-field, and phase-contrast lighting.

The exposures were calculated with a Gossen Lunasix light meter, reading through the reflex viewfinders the light from the ground glass of the cameras. The exposure setting required some experimentation, but once established proved quite reliable.

The types of films employed were kept to a minimum. For black-and-white work with the Rolleiflex, Panatomic X film was used and was processed in D-76 developer. With the Linhof camera, Kodak LS Pan film proved most suitable because of its moderate speed and fine grain. When very high contrast was required, Kodak Contrast Process Ortho film was used. Both of these films were processed in Kodak DK-50 developer.

The color transparencies were all made with the Rolleiflex camera. Kodak Ektachrome X film, with an 80A filter to balance the tungsten lighting, produced good results. However, Agfa CK-120 color film yielded brilliant coloring and was faster. This excellent film is balanced for tungsten lighting (3200 K) and does not therefore require filtering. High Speed Ektachrome, Type B—a fast film also balanced for tungsten lighting—was employed for those specimens that were already intensely colored.

Special filters were necessary to photograph certain stained sections. In black-and-white photography, filters that are the same color as the specimen lighten it, while filters of complementary color darken the subject. The filters most often used for the pictures in this book were the green Kodak Wratten Filter No. 58 and yellow-green No. 11, both useful on red stains; the red No. 25 filter, which darkened the blue and green stains; and the blue No. 47 filter, for yellow and orange preparations. Neutral density filters of different intensities were effective in reducing the amount of light without interfering with the color balance.

BIBLIOGRAPHY

BOOKS FOR THE GENERAL READER

Allen, R. M. *The Microscope*. Princeton, N. J.: Van Nostrand, 1940. (The mechanics and use of the microscope.)

———. "Photomicrography," in *The Encyclopedia of Photography*. New York: Greystone Press, 1967, pp. 2823–38. (A brief review for the amateur photographer.)

Clay, R. S., and Court, T. H. *History of the Microscope*. London: Charles Griffin, 1932. (General historical aspects.)

Corrington, J. D. *Exploring with Your Microscope*. New York: McGraw-Hill, 1957. (An excellent, comprehensive book; highly recommended.)

Croy, O. R. *Creative Photomicrography*. New York: Amphoto, 1968. (Basic techniques for the advanced amateur photographer.)

Dobell, C. *Antony van Leeuwenhoek and His Little Animals*. New York: Russell & Russell, 1958. (An interesting epoch in the development of the microscope.)

Johnson, G. *Hunting with the Microscope*. New York: Sentinel Books, 1963. (A delightful booklet for the amateur explorer.)

MORE ADVANCED AND TECHNICAL TEXTS

Allen, R. M. *Photomicrography.* Princeton, N. J.: Van Nostrand, 1958.

Barron, A. L. E. *Using the Microscope.* London: Chapman and Hall, 1965.

Bergner, J., Gelbke, E., and Mehliss, W. *Practical Photomicrography.* London: Focal Press, 1966.

Chamot, E. M., and Mason, C. W. *Handbook of Chemical Microscopy.* New York: John Wiley, 1956.

Francon, M. *Progress in Microscopy.* Oxford: Pergamon Press, 1961.

Gage, S. H. *The Microscope.* 17th ed. Ithaca, N.Y.: Comstock Publishing Associates, 1941.

Gray, P. *The Use of the Microscope.* New York: McGraw-Hill, 1967.

Hartley, W. G. *Microscopy.* Garden City, N.Y.: Natural History Press, 1964.

Photography Through the Microscope. Kodak Pamphlet P-2. Rochester, N.Y.: Eastman Kodak Co., 1966.

Lawson, D. F. *The Technique of Photomicrography.* New York: Macmillan, 1960.

Olliver, C. W. *The Intelligent Use of the Microscope.* New York: Chemical Publishing Co., 1953.